Once Upon an Earth Science Book

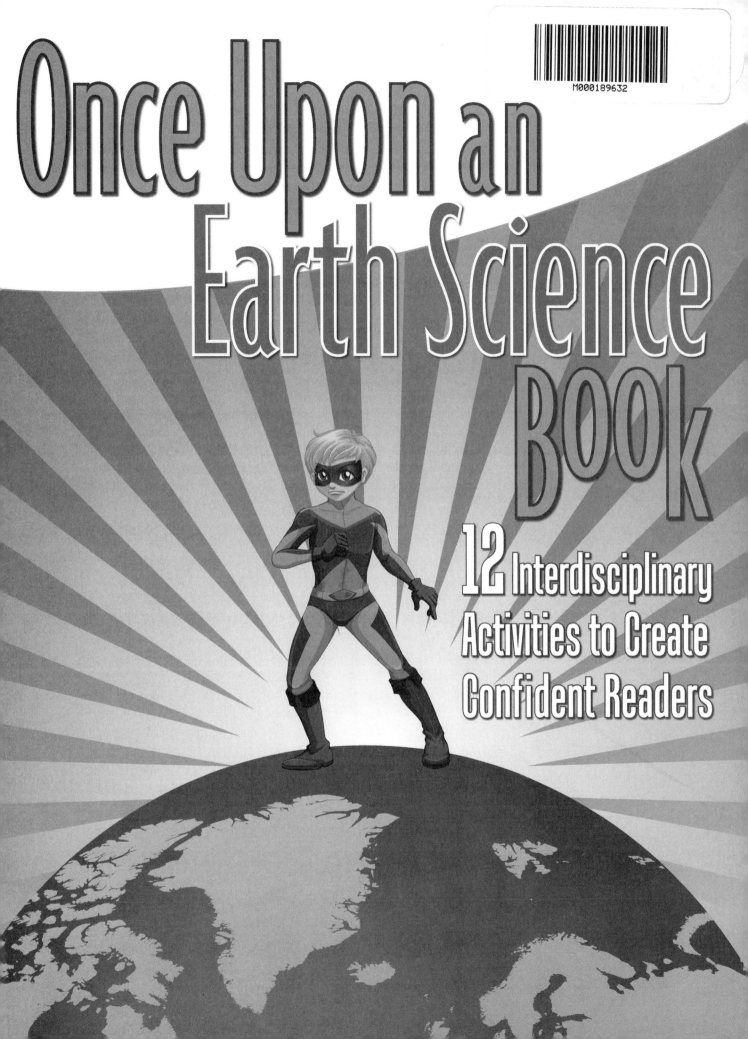

12 Interdisciplinary Activities to Create Confident Readers

Once Upon an Earth Science Book

12 Interdisciplinary Activities to Create Confident Readers

Jodi Wheeler-Toppen

NSTApress

National Science Teachers Association
Arlington, Virginia

National Science Teachers Association

Claire Reinburg, Director
Wendy Rubin, Managing Editor
Rachel Ledbetter, Associate Editor
Amanda O'Brien, Associate Editor
Donna Yudkin, Book Acquisitions Coordinator

ART AND DESIGN
Will Thomas, Jr., Director—Cover and Interior Design
All superhero art on pages 1, 11, 23, 29, 45, 55, 65, 75, 85, 97, 109, 123, 133, 151, and 163 courtesy istockphoto

PRINTING AND PRODUCTION
Catherine Lorrain, Director

NATIONAL SCIENCE TEACHERS ASSOCIATION
David L. Evans, Executive Director
David Beacom, Publisher

1840 Wilson Blvd., Arlington, VA 22201
www.nsta.org/store
For customer service inquiries, please call 800-277-5300.

NSTA is committed to publishing material that promotes the best in inquiry-based science education. However, conditions of actual use may vary, and the safety procedures and practices described in this book are intended to serve only as a guide. Additional precautionary measures may be required. NSTA and the authors do not warrant or represent that the procedures and practices in this book meet any safety code or standard of federal, state, or local regulations. NSTA and the authors disclaim any liability for personal injury or damage to property arising out of or relating to the use of this book, including any of the recommendations, instructions, or materials contained therein.

LIBRARY OF CONGRESS CATALOGING-IN-PUBLICATION DATA
Names: Wheeler-Toppen, Jodi.
Title: Once upon an earth science book : 12 interdisciplinary activities to create confident readers / by Jodi Wheeler-Toppen.
Description: Arlington, VA : National Science Teachers Association, [2016] | Includes bibliographical references.
Identifiers: LCCN 2016007042 (print) | LCCN 2016016494 (ebook) | ISBN 9781941316092 (print) | ISBN 9781941316740 (e-book)
Subjects: LCSH: Earth sciences--Study and teaching (Middle school)--Activity programs. | Earth sciences--Study and teaching (Secondary)--Activity programs. | Geology--Study and teaching (Middle school)--Activity programs. | Geology--Study and teaching (Secondary)--Activity programs. | Oceanography--Study and teaching (Middle school)--Activity programs. | Oceanography--Study and teaching (Secondary)--Activity programs. | Curriculum planning.
Classification: LCC QE28 .W54 2016 (print) | LCC QE28 (ebook) | DDC 550.71/2--dc23
LC record available at https://lccn.loc.gov/2016007042

Contents

Contents

Acknowledgments

For Jon, Natalie, and Zachary

With special thanks to a wonderful team of teachers who field-tested activities from this book:

Donna Budynas
Hutchison School, Memphis, Tennessee

Matt Hackett
Delta Woods Middle School, Blue Springs, Missouri

Jodie Harnden
Sunridge Middle School, Pendleton, Oregon

Michelle Kester
Dike School of the Arts, Cleveland, Ohio

Karen Kraus
Delta Woods Middle School, Blue Springs, Missouri

Judy Strickland
Douglas County Schools, Georgia
National Institutes of Health

Getting Started

In faculty meetings, the principal of my school would periodically exhort those of us who were "subject-area teachers" to contribute to the school-wide emphasis on improving reading. I would return to my classroom and assign pages from the textbook to my students, only to be greeted by moans and groans. After I finally cajoled my groaners into reading, I would ask them questions, but they never seemed to learn much from what they read. Does this sound familiar?

I finally returned to graduate school to find out more about how science teachers could design successful reading lessons for their classes. I compiled what I had learned into a book for life science teachers. I quickly learned I wasn't the only person who had wished for such a book—teachers of other subjects routinely ask me for a book in their area. With the arrival of the *Common Core State Standards (CCSS),* even more teachers are finding themselves struggling to integrate reading, writing, and science.

The good news is that there are many parallels between how people learn science and how they learn to become better readers. In fact, many studies have indicated that integrating reading and science can lead to gains in both areas (e.g., Fang et al. 2008; Kamil and Bernhardt 2004, Morrow et al. 1997; and Radcliffe et al. 2008).

This book is for middle and high school Earth science teachers who are ready to implement successful reading experiences that support science content learning. Each lesson in this book consists of a science activity, a reading about an important Earth science concept (based on a standard from the *Next Generation Science Standards* [*NGSS*]), and a writing activity that asks students to connect what they did with what they read. This book also contains information on teaching reading strategies to help you create a complete reading program in your science class. To begin, let's look at how the processes of science and reading are complementary.

Learning Science: The Learning Cycle

Science teachers know that people learn science best when they anchor their learning in firsthand experiences. This idea has been formalized into the *learning cycle,* a way to organize lessons so that students have a chance to explore a concept before they learn the relevant vocabulary and principles (Lawson 2009). While the 5E learning cycle is often used in science (see Bybee et al. [2006] in the Find Out More box), in this book we will focus on the critical middle three phases because they relate most directly to integrating reading and they form the heart of a learning cycle in both science and reading (Baker 2004).

Exploration. In this phase, students experience a new concept in a concrete way by exploring a specific example. Students may be asked to design an experiment, attempt to solve a problem, or simply make observations. Explorations in this book include designing an experiment to measure the amount of current used by an electrical device, finding ways to destroy a model mountain with erosion, and watching a convection current move through convection tubing. In the exploration, students form an internal framework for the broader ideas that they will encounter during the explanation. This phase also fosters student questions about the topic so they are more interested and engaged during the explanation phase.

Explanation. In this phase, students learn vocabulary and general principles. For example, if students watched a convection current take place during the exploration phase, they would be introduced to the terms *density* and *convection* during the concept introduction phase. They would also learn that the concept they observed during the exploration—that hot liquid has lower density and rises—is a general principle that affects weather and ocean currents. In a learning cycle, concept introductions can come from lectures, discussions, readings, or videos. For the cycles in this book, the explanation is provided through reading.

Concept Application. Finally, students need a chance to apply the new terms and principles themselves in a new situation. The concept application phase of a learning cycle can include doing additional hands-on activities, designing new investigations, making a concept map, or solving a new problem. In this book, the concept application phase for each cycle includes a graphic organizer and writing prompt.

The learning cycle is based in constructivism, a view of learning that holds that students base new knowledge on the understandings they already hold (Lawson 2009). The prior knowledge that students bring to a lesson may be helpful for learning the new material, or students may bring misconceptions that make learning more difficult. The exploration phase can help challenge students' misconceptions so they are ready to restructure their understanding.

The exploration also fills in gaps that may be present in a student's prior knowledge. For example, if a student's main exposure to rocks has been gravel, he or she will find it difficult to understand that different rocks have different properties. Observing and conducting tests on different types of rock during the exploration phase can provide the necessary background for learning the rock cycle.

Developing Reading Skills: What Reading Teachers Know

It turns out that prior knowledge is important in reading as well (Rosenblatt 1994). Students often do not understand a text because they lack information that the author of the text assumes they know (Fielding and Pearson 1994). To help students access useful prior knowledge and develop new background that they will need for a text, reading educators use a plan for reading instruction that is very similar to the learning cycle (Robb 2000).

Pre-Reading. During pre-reading, reading teachers ask questions to help students think about what they already know that will be important for understanding the text. For example, if students are going to read about a volcano, the teacher may ask what students know about volcanoes or have them draw a picture of a volcano. Reading teachers call this activating prior knowledge. In some cases, reading teachers need to introduce new background information that students will need for the reading, such as sharing about the time period in which a story is set or showing students an example of an object that plays a key role in the text.

FIND OUT MORE

If learning cycles are new to you, see the following:

- Bybee, R. W., J. A. Taylor, A. Gardner, P. Van Scotter, J. C. Powell, A. Westbrook, and N. Landes. 2006. *The BSCS 5E Instructional Model: Origins, effectiveness, and applications.* Colorado Springs: BSCS. *www.bscs.org/curriculumdevelopment/features/bscs5es.html.*
- Chapters 5 and 6 in Lawson, A. 2009. *Teaching inquiry science in middle and secondary schools.* Thousand Oaks, CA: SAGE Publications.

For an excellent overview of how the learning cycle fits with what we know about how people learn, see the introduction to

- Moyer, R., J. K. Hatchett, and S. A. Everett. 2007. *Teaching science as investigations: Modeling inquiry through learning cycle lessons.* Upper Saddle River, NJ: Pearson Education.

FIND OUT MORE

Read more about prior knowledge and misconceptions in the following:

- Introduction to National Research Council (NRC). 2005. *How students learn: Science in the classroom.* Washington, DC: National Academies Press.
- Any volume of Page Keeley's *Uncovering Student Ideas in Science* series (NSTA Press).

Reading. This step includes the obvious task of reading the passage, but reading teachers may expand it by asking students to use specific strategies to monitor their comprehension as they read.

Post-Reading. After reading, students are led in reflecting on what they learned and applying this knowledge in a new situation. Reading teachers often have students summarize main ideas, create concept maps, or do projects based on what they read.

A Natural Fit

You can see how the work of science teachers and reading teachers fits well together. An exploration can serve as a pre-reading activity by generating background knowledge that supports new learning. It also provides an authentic purpose for reading. No more reading just to answer questions at the end of the chapter—now students can read to answer real questions that they have developed from experiencing science firsthand. For the next phase, reading provides an excellent source for the concept introduction. Reading also models the work of real scientists, who usually read a great deal in the process of developing and interpreting experiments. Finally, both models call for an activity in which students apply their developing knowledge.

One difference in the two models is that reading educators focus more of their attention on what takes place during the actual reading. They know that good readers monitor their comprehension as they read. Reading teachers help students pay attention to whether they understand and teach strategies that improve comprehension and memory. Science educators can also teach these strategies, and once again, reading research is on our side—research has shown that students learn reading strategies best if the strategies are incorporated into meaningful reading opportunities (Baker 2004; Fielding and Pearson 1994).

The Larger Goal

It is important to remember, when teaching reading in science, that our goal is bigger than reading comprehension. We do not want our students to make sense of just one bit of text; we want them to draw from several texts, their prior experiences, conversations, labs, lecture, and media and video and to assemble it all into an understanding of the topic. We want them to read as an act of "knowledge building" (Cervetti and Hiebert 2015).

Making the Most of This Book

One way to use this book would involve selecting certain lessons that suit your curriculum and using them à la carte. However, this book can also be used as a reading development program by spacing the lessons throughout the year and including the following components that are provided with each chapter:

Strategy Introduction. Each lesson in this book includes a specific reading comprehension strategy that you can introduce before students read. Teaching this strategy does not need to take a great deal of direct instruction; much of the strategy learning will take place during individual practice and small-group interactions. However, explicitly addressing reading strategies can help students learn to take control of their reading (Baker 1991; Radcliffe et al. 2008; Spence, Yore, and Williams 1999).

Reading Groups. Working in reading groups can be a powerful tool for improving comprehension (Rosenshine and Meister 1994; Wheeler-Toppen 2006). By working together to read a passage, students can fill in gaps in prior knowledge and model reading strategies for each other (Robb 2000). Talking about what they have read causes students to observe how successfully they are comprehending it, and over time, that awareness can lead to self-regulation even outside of the group (Baker 2004).

There are a number of ways to organize successful reading groups, and in the next chapter I will introduce one such procedure that is simple enough to be implemented in most science classrooms.

Journaling. It takes time, practice, and reflection for a new reading strategy to become a stable part of a student's repertoire. Reading teachers often conduct individual reading conferences with their students to reinforce new strategies, monitor progress, and help students reflect on their development as readers (Allen 1995; Robb 2000; Schoenbach et al. 1999). Science teachers rarely have time to implement individual reading conferences in their classes. Therefore, this book includes journal questions to encourage students to internalize the strategies introduced in class. These questions are designed to help students plan ways to use a strategy, practice a strategy by writing an example situation, or evaluate their own use of a strategy. You can increase the value of the journaling activity by periodically responding to entries in your students' journals.

Assessing Student Learning

Assessment is a critical piece of any teaching endeavor. The assessment exercises in this book are based on the idea that assessments should give you feedback on what your students are learning and also serve as learning opportunities for your students (NRC 1996). Therefore, each activity, in addition to being a learning tool, is designed to provide information about how well your students understand the lesson. These assessments fall into four general categories:

The Big Question. The Big Question is a reading comprehension check found at the end of each reading selection. The question can be answered in a fairly straightforward manner from the information in the text but cannot be answered simply by skimming or quoting a section of text. Answers to the Big Question are generally short, and if students work in reading groups, each group may submit just one answer. You should be able to scan the answers relatively quickly to determine if students grasped the main ideas of the text.

Thinking Visually. In science, many ideas are best represented using diagrams, concept maps, and other graphic organizers. These tools are important pieces of text that students often overlook. Learning to "read" visual representations is a significant part of reading and learning science. Likewise, having students organize their own thinking visually is a powerful tool (Fisher 2002). Therefore, each chapter in this book includes an activity that asks students to either reflect on an existing visual representation of key ideas from the chapter or use a visual framework to organize their new knowledge.

Writing Prompt. The last activity in each chapter is a writing activity that asks students to integrate what they learned from the exploration and the text. Each activity focuses on a key idea. That idea will be listed below the writing prompt. You can use the rubric in Table 1.1, along with this focus point, to assess your students' responses to the writing prompt. Because these are open-ended questions, they provide an especially good opportunity to watch for misconceptions that your students may have developed.

Claims and Evidence. One science skill highlighted in this book is the ability to use claims and evidence. To this end, six of the explorations ask students to make a claim and support it with evidence. This ability should be assessed as well because what we choose to assess communicates to

> **FIND OUT MORE**
> For more information on helping students "read" visual representations of concepts, consider the following resource:
> • Vasquez, J. A., M. W. Comer, and F. Troutman. 2010. *Developing visual literacy in science, K–8.* Arlington, VA: NSTA Press.

Table 1.1. Rubric for Evaluating Responses to Writing Prompts

Question	Completely	Partially	Incorrectly or Insufficiently
Does the response correctly describe the key idea?			
Is the response supported by details from the reading or investigation?			
What misconceptions are present in the response?			

students what we think is important. Information about assessing the claims and evidence aspects of the explorations can be found in Chapter 3.

Additionally, you will want to assess how students are developing as readers. This is challenging because much of what we teach students to do happens invisibly as they read. However, you can listen to what students say in their reading groups and write in their reading journals. You can also give students the self-evaluation found in Chapter 2 several times during the year to look for growth.

Remember that assessment should affect what you do next with your students (Black 2003). If you find that students are struggling with the Big Question or graphic organizer, you may need to talk about the text with your class. If you encounter misconceptions in the written responses, you need to address them. If students are struggling with claims and evidence, stop and take a day to try one of the activities from Chapter 3 that will help

IMPORTANT SAFETY INFORMATION

With hands-on, process-, and inquiry-based activities and investigations, the teaching and learning of science today can be both effective and exciting. The challenge to securing this success needs to be met by addressing potential safety issues by conducting hazard assessments and using appropriate engineering controls (ventilation, fume hoods, fire extinguishers, showers, etc.), administrative procedures and safety operating procedures and use of appropriate personal protective equipment (indirectly vented chemical-splash goggles meeting the ANSI Z87.1 standard, chemical resistant aprons, nonlatex gloves, etc.). Teachers can make it safer for students and themselves by adopting, implementing, and enforcing legal safety standards and better professional safety practices in the science classroom and laboratory. Throughout this book, Safety Alerts are provided for investigations and activities and need to be adopted and enforced in an effort to provide for a safer learning and teaching experience. Teachers should also review and follow local polices and protocols used within their school district and school (e.g., chemical hygiene plan, Board of Education safety policies).

Additional applicable standard operating procedures can be found in the National Science Teachers Association's *Safety in the Science Classroom, Laboratory, or Field Sites* document (*www.nsta.org/docs/ SafetyInTheScienceClassroomLabAndField.pdf*). Students should be required to review the document or one similar to it under the direction of the teacher. Both the student and the parent or guardian should then sign the document acknowledging procedures that must be followed for a safer working and learning experience.

Disclaimer: The safety precautions of each activity are based, in part, on use of the recommended materials and instructions, legal safety standards, and better professional practices. Selection of alternative materials or procedures for these activities may jeopardize the level of safety and therefore is done at the user's own risk.

students develop that skill. Conversely, if students are successful with these activities, you can move on with the confidence that they are ready for new topics.

Get Started!

With this book, you have everything you need to boost your students' science and reading skills. Start by learning about the strategies you need for the book in Chapters 2 and 3. Then dive into the 12 content chapters. As you and your students work through these lessons together, you will be able to watch their confidence as readers—and your confidence as a reading educator—grow. So what are you waiting for? Let's get started!

References

Allen, J. 1995. *It's never too late: Leading adolescents to lifelong literacy.* Portsmouth, NH: Heinemann.

Baker, L. 1991. Metacognition, reading, and science education. In *Science learning: Processes and applications*, ed. C. M. Santa and D. E. Alvermann, 2–13. Newark, DE: International Reading Association.

Baker, L. 2004. Reading comprehension and science inquiry: Metacognitive connections. In *Crossing borders in literacy and science instruction*, ed. E. W. Saul, 239–257. Newark, DE: International Reading Association.

Black, P. 2003. The importance of everyday assessment. In *Everyday assessment in the science classroom*, ed. J. M. Atkin and J. E. Coffey, 1–11. Arlington, VA: NSTA Press.

Cervetti, G. N., and E. H. Hiebert. 2015. Knowledge, literacy, and the *Common Core. Language Arts* 92 (4): 256–269.

Fang, Z., L. Lamme, R. Pringle, J. Patrick, J. Sanders, C. Zmach, S. Charbonnet, and M. Henkel. 2008. Integrating reading into middle school science: What we did, found, and learned. *International Journal of Science Education* 30 (15): 2067–2089.

Fielding, L. G., and P. D. Pearson. 1994. Reading comprehension: What works. *Educational Leadership* 51 (5): 62–68.

Fisher, K. M. 2002. Overview of knowledge mapping. In *Mapping biology knowledge*, ed. K. M. Fisher, J. H. Wandersee, and D. E. Moody, 5–24. New York: Kluwer Academic Publishers.

Kamil, M. L., and E. Bernhardt. 2004. The science of reading and the reading of science: Successes, failures, and promises in the search for prerequisite reading skills for science. In *Crossing borders in literacy and science instruction*, ed. E. W. Saul, 123–139. Newark, DE: IRA.

Lawson, A. 2009. *Teaching inquiry science in middle and secondary schools.* Thousand Oaks, CA: SAGE Publications.

Morrow, L. M., M. Pressley, J. K. Smith, and M. Smith. 1997. The effect of a literature-based program integrated into literacy and science instruction with children from diverse backgrounds. *Reading Research Quarterly* 32 (1): 54–76.

National Research Council (NRC). 1996. *National science education standards.* Washington, DC: National Academies Press.

Radcliffe, R., D. Caverly, J. Hand, and D. Franke. 2008. Improving reading in a middle school science classroom. *Journal of Adolescent and Adult Literacy* 51 (5): 398–408.

Robb, L. 2000. *Teaching reading in middle school: A strategic approach to reading that improves comprehension and thinking.* New York: Scholastic Professional Books.

Rosenblatt, L. 1994. The transactional theory of reading and writing. In *Theoretical models and processes of reading,* ed. R. Ruddell, M. Ruddell, and H. Singers, 1057–1092. 4th ed. Newark, DE: International Reading Association.

Rosenshine, B., and C. Meister. 1994. Reciprocal teaching: A review of the research. *Review of Educational Research* 64 (4): 479–530.

Schoenbach, R., C. Greenleaf, C. Cziko, and L. Hurwitz. 1999. *Reading for understanding: A guide to improving reading in middle and high school classrooms.* New York: Jossey-Bass.

Spence, D. J., L. D. Yore, and R. L. Williams. 1999. The effects of explicit science reading instruction on selected grade seven students' metacognition and comprehension of specific science text. *Journal of Elementary Science Education* 11 (2): 15–30.

Wheeler-Toppen, J. 2006. Reading as investigation: Using reading to support and extend inquiry in science classrooms. PhD diss., University of Georgia, Athens.

The Reading Strategies

The belief that reading is essentially a process of saying the words rather than actively constructing meaning from texts is widespread among many students. For instance, one of the students we interviewed looked surprised when he was asked to describe the topic discussed in a section of text he had just read. "I don't know what it was about," he answered, with no sense of irony. "I was busy reading. I wasn't paying attention." (Schoenbach et al. 1999, p. 6)

What Is Reading?

At lunch, my colleagues and I would periodically bemoan how our students "can't read." But what did we really mean by that? Certainly, most of our students, even those scoring well below grade level on reading tests, could pronounce the words on the page of a simple book. Some of them even enjoyed reading novels for fun. But they seemed completely unable to make sense of their science textbooks or other school books.

Part of the problem lay in the way they thought about reading. Like the young man in the example above, they believed that reading consisted of calling out the words on the page. Good readers, they assumed, automatically understood all of those words, and their own failure to do so simply reinforced students' beliefs that they didn't read well. Reading strategies can be important for helping students improve their reading, but students need something more.

They need to begin to view reading as an active search for meaning that is within their control. We can change how our students think about reading through the way we talk about reading in our classrooms.

Starting the Conversation

The first step is to create a classroom culture in which students feel safe exploring new ways of thinking about reading. You can begin by simply stating aloud that reading can be difficult, even for good readers. You can share your own stories about encountering words, phrases, or books that were hard for you to understand. Most important, you should make it clear that you will not tolerate students teasing each other about reading struggles.

Next, students need a chance to see what experienced readers do as they read. Good readers constantly monitor their comprehension and notice if they do not understand what they read. They often have an ongoing conversation in their head in which they compare what they are reading with what they already know (and sometimes argue with the text if they disagree). When they do not understand, or they find inconsistencies with their prior knowledge, they use problem-solving strategies to make sense of the text or resolve the inconsistencies. All of these things are hidden from someone watching, but as teachers we can make them visible.

Think-Alouds. One way to make the invisible processes of reading visible is to talk about what we are thinking as we read (Baker 2004; Kucan and Beck 1997). This is called a think-aloud. For example, you might read this section from the reading selection in Chapter 13:

> *Our solar system began in a cloud of dust and gas that was probably leftover from a star that had exploded. At first, there was just a spinning disc of debris. But this matter had gravity, and the pieces were attracted to each other. Most of it clumped together in the middle and formed our star, the Sun. But at the same time, smaller clumps were forming in orbit around the Sun. These clumps would eventually become the planets of our solar system.*

To use this in a think-aloud, you would insert your own thoughts as you read out loud, so it might sound something like this:

> *Our solar system began in a cloud of dust and gas that was probably leftover from a star that had exploded.* So, there was probably a star here before our solar system? I didn't know that. *At first, there was just a spinning disc of debris.* I know debris is like leftover trash, so

this must be leftovers from the star. *But this matter had gravity, and the pieces were attracted to each other. Most of it clumped together in the middle and formed our star, the Sun.* That says "most of it," so I bet it's about to tell me what happened to the rest of it. *But at the same time, smaller clumps were forming in orbit around the Sun. These clumps would eventually become the planets of our solar system.* Aha, I was right.

This allows struggling readers to "see" how strong readers approach difficult reading passages. You can use a think-aloud to demonstrate specific strategies to your class or when you are helping a student or small group figure out a confusing passage.

Peer Conversation. Students can also show invisible aspects of reading to each other. In reading groups (see discussion on p. 14), students share with each other how they made sense of the text. One student might read the above passage and ask the group, "What's this 'debris' word?" Another student might answer by saying that she's heard trash called debris. A third student might add that an earlier sentence mentions "a cloud of dust and gas that was probably leftover from a star." You may be skeptical that your students would be able to have these discussions with each other. You will be surprised. As students get used to working in groups, and as an atmosphere of trust develops, even weak readers become comfortable asking for help and sharing what they think as they read.

Overarching Strategies

Each lesson in this book can be used to introduce students to one or more specific reading strategies. These strategies are important; they represent ways that good readers solve specific reading problems. However, keep in mind that learning specific strategies is not the ultimate goal. We want students to begin to approach reading as an active search for meaning (Cervetti and Hiebert 2015; Loxterman, Beck, and McKeown 1994). The following strategies are only a means to that end.

The first two strategies introduced here, comprehension coding and reading groups, are intended to be used throughout all of the lessons. They address the primary issues of comprehension monitoring and problem solving.

Comprehension Coding. In comprehension coding, students mark codes to indicate what they are thinking as they read. I recommend introducing the following codes:

- ! This is important.
- ✓ I knew that.
- × This is different from what I thought.
- ? I don't understand.

Over time, students may develop their own coding systems that meet their particular needs. Indeed, you may notice that you do something similar yourself. You may underline important information you want to remember or jot questions in the margins of books. This strategy is intended to mimic that sort of behavior on the part of good readers and encourage students to monitor their comprehension as they read.

Reading Groups. Working in reading groups can be a powerful tool for improving comprehension (e.g., Rosenshine and Meister 1994; Wheeler-Toppen 2006). There are a number of ways to organize reading groups; however, I recommend the following simple procedure for the activities in this book.

In this procedure, each reading group has three students with specific jobs. The *leader* guides the group through the procedure listed on the board or a sheet of paper (see Figure 2.1). The *emergency manager* raises an orange flag when the group needs help from the teacher. This flag can be a folded piece of construction paper that is propped up or a plastic cup that is placed on the desk when help is needed. The *interpreter* records the group's answer to the Big Question, a general comprehension check that follows the reading (see Chapter 1 for further explanation).

Figure 2.1. Reading Group Procedure

1. Everyone reads the first section quietly and marks !, ✓, ×, and ? while reading.
2. The leader asks each member of the group to share anything that was confusing (marked ? or ×).
3. The group should try to figure out what the confusing word, sentence, or idea means. If the group cannot make sense of the confusing word, sentence, or idea, the emergency manager should raise the flag to get help from the teacher.
4. Repeat steps 1–4 for Section 2.
5. Repeat steps 1–4 for Section 3.
6. The group should work together to answer the Big Question. The interpreter will write the group's answer to turn in to the teacher.

This group procedure calls for reading passages to be broken into three sections. You will find that the readings in this book are divided into sections by short black lines so they can be used with or without this procedure.

Your role as the teacher is important during this process. Initially, you will need to monitor students closely to ensure that they really do follow the procedure. When students raise their flags, listen to their comprehension difficulty. What have they tried so far? Can you model a strategy for making sense of the text? Is there a piece of background knowledge or a word meaning that you need to provide?

Listen in on groups that are not having trouble as well. Encourage students to share their strategies. For example, if one student tells another what a sentence means, ask, "How did you know that?" By participating in the groups with students, you show that even teachers have to think carefully about what they read.

As with any classroom procedure, this one takes practice. For the first few sessions, students will have to focus as much on what to do as on what they are reading. They will also need to see that you are serious about requiring them to follow the procedure. After two or three sessions, however, students will begin to follow the procedure automatically and be able to focus more on content. The time invested in learning the process will be well worth it, as students' reading skills and confidence improve.

Problem-Solving Strategies

The rest of the strategies in this book are designed to help students solve specific comprehension problems or learn something about how science texts are organized. Each chapter will describe how to introduce one of the following strategies that would be appropriate to use while reading the article for that chapter. Before giving the article to your class, read it yourself and identify places the strategy would be useful. This will help you guide your students' reading.

Keep in mind that students need practice to master any strategy. For this reason, it will be important for you to monitor students closely the first time they use a strategy. In this book, several strategies appear in two different lessons to reinforce their use. You may also want to follow up with readings from your textbook or other sources to allow them to practice the strategies further.

> **TEACHING NOTE**
> Some teachers have difficulty using comprehension coding with their students because their school restricts the number of copies they can make. If you are in this position, consider the following ideas:
> - Talk to your administrator about the problem. Most administrators are interested in supporting attempts to improve reading. They may be willing to allow you extra copies for this purpose.
> - Consider printing a class set of readings and having them laminated or slipped into clear page protectors. Students can code using overhead projector pens and then wipe off their marks for the next class. These class sets can be used for several years.
> - To use this strategy with textbooks, you can give students strips cut from sticky notes to mark sentences as they read.

> **TEACHING NOTE**
> Science textbooks are examples of expository text, a type of writing that describes or explains a concept. Expository text stands in contrast to narrative text, which tells a story. Many science magazine articles use the story of a specific situation to introduce science concepts, so they include both expository and narrative text. I have written some of the articles in this book to reflect textbook-style writing and some to reflect magazine-style writing.

Finding the Meaning of New Words. Specialized vocabulary is a key feature of science texts (Fang 2006; Holliday 1991). As students read, they are continually introduced to new terms. Struggling readers often miss the definitions of the words when they are introduced because they don't recognize the cues that a definition is being given.

Most students will have been introduced to the strategy of using context clues to find the meaning of new words. As a general reading strategy, using context clues means looking at the surrounding text to figure out a likely meaning of the unfamiliar word. Although many students know they should "use context clues," they often do not use the strategy successfully. Chapters 5 and 7 help students practice this skill by focusing on some of the most common ways that new definitions are presented in science text (see Table 2.1). Students can learn to look for the clues in the sentence before and after a new word is used for the first time.

Note that sometimes the text does not provide sufficient context clues for students to figure out the meaning of a word, especially for nonscience words or vocabulary that are not the focus of the reading. These words constitute background knowledge that the writer expects students to have already. One advantage to having students in reading groups is that they can help each other with these words. Alternatively, if you can identify words that may cause problems, you can teach the words before giving students the text.

> **TEACHING NOTE**
>
> Sending students to the dictionary, or even the glossary in the back of a book, is not a particularly productive way to help them with unknown words. Dictionary definitions can be more confusing than the original text and include additional words that students do not know. Furthermore, the task of looking up the word interrupts a student's reading process. It is much less disruptive to simply tell students the meaning of a word if it cannot be figured out from the context.

Table 2.1. Common Ways That Texts Introduce New Words

Example	Explanation
Soil can be washed away by runoff. Runoff is the water that collects and moves across the ground during a rainstorm.	The sentence after the term provides a definition.
Water that moves across the ground during a rainstorm, called runoff, …	The new term is signaled with the word *called*.
Soil can be washed away by runoff, which is rainwater that runs across the ground.	The definition is signaled with the phrase *which means* or *which is*.
Runoff, or water that runs across the ground in a rainstorm, …	The word *or* after a comma indicates that the word and phrase mean the same thing.
Soil can be washed away by runoff. This movement of rainwater across the ground …	This is the trickiest situation. The text doesn't directly say what the word means, but implies it by using the word and the definition close together.

Previewing Diagrams and Illustrations. When students read science books, they often ignore the diagrams and illustrations (Wheeler-Toppen 2006). By doing this, they miss out on excellent reading support and, often, key information (Holliday 1991; Vasquez, Comer, and Troutman 2010). Diagrams and illustrations can clarify points that are hard to explain in words. They may provide useful background knowledge or give examples. Pictures can be especially helpful for struggling readers because they provide hints about what the text will say.

Previewing diagrams and illustrations can serve several functions. When you, as the teacher, lead the preview, you can use it to call attention to important points that the text will make. You can also use it to help students draw on their prior knowledge by having them look for things that they recognize in the pictures. Previewing can also be a useful tool for helping students make predictions before they read. When students begin reading, they will attempt to find out if their predictions were correct. Chapters 13 and 14 take advantage of all of these possibilities. Ultimately, though, we want students to be able to use diagrams and illustrations without teacher guidance. To that end, the most important point to make with your students is that diagrams and illustrations are a critical part of science texts. Students should get in the habit of studying diagrams and illustrations, preferably before they read.

Text Signals. Expository text often includes key words that can signal what the reader can expect to find in the next few phrases or sentences (Schoenbach et al. 1999). Some teachers have compared these text signals to traffic signs. When readers see these words or phrases, they should slow down and notice the information that follows. Chapters 11 and 12 focus on two common groups of signal words. Chapter 11 looks at signals for cause and effect, and Chapter 12 presents signals for comparing and contrasting (see Table 2.2, p. 18).

Although these are the only signal words that are introduced formally in this book, there are many other types. For example, words such as *following, previously,* and *during* indicate that a timeline is being given. The presence of a question in a text generally indicates that an answer will be given. Signals such as *like, including, such as, for instance,* and *e.g.* tell the reader that examples are being given. You may find opportunities to introduce these or other text signals in your conversations with students about reading.

Table 2.2. Text Signals Introduced in Chapters 11 and 12

Cause and Effect	*consequently, so, as a result, therefore, for this reason, thus, because, hence, in response to* (*since* often indicates cause and effect)
Comparisons	*similarly, in the same way, just like, just as, likewise,* and *also*
Contrasts	*however, in contrast, on the other hand, conversely,* and *whereas, alternatively, instead* (*but, yet,* and *while* sometimes indicate a contrast)

Chunking. Science texts often have a lot of information crammed into each sentence (Fang 2006). Consider the following basic sentence about granite:

Nonporous, igneous rocks, such as granite, are often used as building materials.

To understand the subject of this sentence—nonporous, igneous rocks—the reader has to understand that the sentence is about rocks; that these rocks are nonporous, which requires remembering that *non* means not; and that they are igneous, and since *igneous* is probably a relatively new word to the reader, he or she must think through where igneous rocks come from. And this is all before they even get to the main point of the sentence.

The example sentence also includes the phrase *such as granite,* which is stuck in the middle of the main thought. This type of phrase, which is common in science writing, is called the interruption construction (Fang 2006) because it interrupts the flow of the sentence. If the interruption is long, it can be especially confusing for struggling readers.

Experienced readers intuitively break sentences like the one above into chunks and think about each chunk individually. Struggling readers may try to understand the whole sentence at once (Schoenbach et al. 1999). The reading strategy of chunking helps students break sentences into separate ideas. Chapters 6 and 8 introduce students to chunking by showing them that they can stop to think as they read, even when there is no period or comma to signal a pause.

Talk Your Way Through It. Sometimes students will tell me that they read their book, but they can't remember what they read. They may be working so hard to understand individual sentences that they fail to grasp what the text as a whole means. Another issue, especially with science, is that the text may contain so much information that it is difficult to remember. Talk your way through it is essentially a summarizing strategy, but I find that students are less intimidated by the phrase "talk your way through it" than they are by the idea of doing something as formal sounding as "summarize."

In the talk your way through it strategy, students stop throughout the text to rehearse the information they have just read. They may need to use a sentence starter such as, "This section is saying that …" or "What I understand so far is …" If they have difficulty stating it in their own words, then they know they need to reread the section for clarification. Chapters 9 and 10 provide a modified reading group procedure that includes taking time to talk through the dense material presented in those chapters.

Evaluating Persuasive Science Writing. The final strategy in this book deals with the overlap of science and public policy. While much argument writing in science deals with interpreting evidence in the process of making explanations, science is also invoked in deciding public policy. Chapter 15 introduces students to some of the questions that readers should ask of policy arguments including the following:

- What evidence does the writer give to support his or her opinion?
- What are the sources of any data cited, and are they reliable and timely?
- How do the scientific explanations given relate to other science information I know?
- What counterarguments are presented in the discussion?
- What counterarguments might the writer be missing?

Chapter 15 is only an introduction to this type of critical reading, but our students need to consider these kinds of questions as they become expert science readers.

Using Self-Assessment to Monitor Strategy Development

As discussed in Chapter 1, reading teachers often use individual and small-group conferences to assess how students are developing as readers. Such conferences are beyond the scope of what most science teachers are able to incorporate into their classes. Therefore, you can monitor your students' strategy use and reading confidence by using a periodic selfassessment. To conduct the assessment, have students answer the following questions in their reading journals (adapted from Robb 2000):

- What do you do before reading to get ready to learn?
- While reading, what do you do if you come to a word or section that you do not understand?
- How do you help yourself remember the details of your reading?
- What would you like to do better as a reader?

Then give students a slip of paper with the chart in Table 2.3 to complete and tape into their journals.

Give the assessment first at the beginning of the year, before you start to teach the strategies. Read over your students' answers, as you may want to make changes to your instruction based on what they say. Give the assessment again midyear and at the end of the year, and encourage students to look over their previous responses and note how they have improved.

Table 2.3. Self-Assessment of Strategy Use

How Often Do I ...?	A Lot	Sometimes	Never
Use codes (such as ✓, +, ?, and ×) to mark what I'm thinking as I read			
Use the information around a new word to figure out what it means			
Study the diagrams and illustrations before reading			
Use text signals to recognize causes and effects			
Use text signals to recognize comparisons and contrasts			
Chunk difficult sentences into smaller pieces to read			

References

Baker, L. 2004. Reading comprehension and science inquiry: Metacognitive connections. In *Crossing borders in literacy and science instruction*, ed. E. W. Saul, 239–257. Newark, DE: IRA.

Cervetti, G. N., and E. H. Hiebert. 2015. Knowledge, literacy, and the *Common Core. Language Arts* 92 (4): 256–269.

Fang, Z. 2006. The language demands of science reading in middle school. *International Journal of Science Education* 28 (5): 491–520.

Holliday, W. G. 1991. Helping students learn effectively from science text. In *Science learning: Processes and applications,* ed. C. M. Santa and D. E. Alvermann, 38–47. Newark, DE: International Reading Association.

Kucan, L., and I. L. Beck. 1997. Thinking aloud and reading comprehension research: Inquiry, instruction, and social interaction. *Review of Educational Research* 67 (3): 271–299.

Loxterman, J. A., I. L. Beck, and M. G. McKeown. 1994. The effects of thinking aloud during reading on students' comprehension of more or less coherent text. *Reading Research Quarterly* 29 (4): 352–367.

Robb, L. 2000. *Teaching reading in middle school: A strategic approach to reading that improves comprehension and thinking.* New York: Scholastic Professional Books.

Rosenshine, B., and C. Meister. 1994. Reciprocal teaching: A review of the research. *Review of Educational Research* 64 (4): 479–530.

Schoenbach, R., C. Greenleaf, C. Cziko, and L. Hurwitz. 1999. *Reading for understanding: A guide to improving reading in middle and high school classrooms.* New York: Jossey-Bass.

Vasquez, J. A., M. W. Comer, and F. Troutman. 2010. *Developing visual literacy in science, K–8.* Arlington, VA: NSTA Press.

Wheeler-Toppen, J. 2006. Reading as investigation: Using reading to support and extend inquiry in science classrooms. PhD diss., Univ. of Georgia, Athens.

How Do You Know That?

Helping Students With Claims, Evidence, and Reasoning

When I wrote the first book in this series (*Once Upon a Life Science Book*), many teachers were not familiar with the concept of scientific argumentation—the process of making claims and supporting them with evidence. Since that time, both the *Next Generation Science Standards* (*NGSS*) and the *Common Core State Standards* (*CCSS*) have emphasized the importance of argumentation. Let us take a moment to examine what those documents have to say and to consider why argumentation is of value for the science student.

Next Generation Science Standards

In the *NGSS*, Engaging in Argument From Evidence is one of the eight science and engineering practices that students are to learn. *A Framework for K–12 Science Education* (*Framework*; NRC 2012), on which the *NGSS* were based, states the following:

> The study of science and engineering should produce a sense of the process of argument necessary for advancing and defending a new idea or an explanation of a phenomenon and the norms for conducting such arguments. In that spirit, students should argue for the explanations they construct, defend their interpretations of the associated data, and advocate for the designs they propose. (p. 73)

Here, the *Framework* alludes to one of the most important reasons that science students should practice argumentation, namely, that *this is what scientists do.* Scientists use their research and observations to make a claim about the natural world. Then they describe the evidence that led them to make that claim and explain how they used reason when looking at evidence to make their claim. The evidence may come from a variety of sources, including controlled experiments in which only one variable is adjusted at a time (your classic schoolroom "scientific method"), analysis of patterns within data sets, physical or computer models, and comparisons between historical evidence and modern processes.

Common Core State Standards

Argumentation is particularly relevant to this book because making claims and supporting them with evidence are intricately tied to reading and writing. Although scientists sometimes present claims and evidence orally (in discussions with colleagues, at conferences, etc.), the written words contained in science journals are ultimately the record of science knowledge. Writing about claims and evidence has content value as well. When students write and talk about the claims and evidence they have made, their content learning improves (Norton-Meier et al. 2008).

The *CCSS* address the kinds of arguments that students should be able to read and write. They provide separate expectations for reading and writing in history/social studies versus science and technical subjects starting in sixth grade. It is worth studying the complete description of what students are expected to be able to do when making written claims:

- **CCSS.ELA-Literacy.WHST.6-8.1.** Write arguments focused on discipline-specific content.
- **CCSS.ELA-Literacy.WHST.6-8.1.a.** Introduce claim(s) about a topic or issue, acknowledge and distinguish the claim(s) from alternate or opposing claims, and organize the reasons and evidence logically.
- **CCSS.ELA-Literacy.WHST.6-8.1.b.** Support claim(s) with logical reasoning and relevant, accurate data and evidence that demonstrate an understanding of the topic or text, using credible sources.
- **CCSS.ELA-Literacy.WHST.6-8.1.c.** Use words, phrases, and clauses to create cohesion and clarify the relationships among claim(s), counterclaims, reasons, and evidence.

- **CCSS.ELA-Literacy.WHST.6-8.1.d.** Establish and maintain a formal style.
- **CCSS.ELA-Literacy.WHST.6-8.1.e.** Provide a concluding statement or section that follows from and supports the argument presented (NGAC and CCSSO 2010).

You will notice that these requirements fit neatly with the expectations laid out in the *NGSS*. Notice also that these standards are intended for students of social studies as well as science and technical subjects. Just as scientists make claims and support them with evidence, so do the majority of academics in all fields. Claims and evidence are the language of logical thought.

The difference between arguments as they are presented in science versus other subjects has to do with what passes as a claim and what passes as evidence for that claim. The "Introduction to the Language Arts Standards" of the *CCSS* indicates that students should "know that different disciplines call for different types of evidence (e.g., documentary evidence in history, experimental evidence in science)" (NGAC and CCSSO 2010).

One way to illustrate these differences is to compare scientific argumentation with courtroom argumentation. Like scientists, lawyers make claims and support them with evidence. However, a lawyer's main interest is to make claims that help his or her client. Counterevidence, or evidence that would hurt a client's claim, is downplayed. In science, however, the goal is to find the claim that best explains all existing evidence, so looking for counterevidence is an important part of making a scientific claim (Lawson 2003).

Scientific Argumentation

Much has been written on helping students with claims and evidence (see Find Out More, p. 27), and I will not try to replicate all of that information here. However, you may find it useful to consider some of the rhetorical underpinnings of argumentation.

In the 1950s, the philosopher Stephen Toulmin identified six major features of scientific arguments, four of which we will examine here: claims, evidence, warrants, and backings (Toulmin 1958). According to Toulmin's model, in a scientific argument evidence is presented to support a claim. Warrants then explain why this evidence provides legitimate support.

Backings—essentially, evidence that a warrant is valid—are sometimes used as well. Consider this everyday example. I came home the other day and found my shoe lying in pieces in the living room and my dog hiding under the bed. Using Toulmin's model, I make the claim that my dog chewed up my shoe, as follows in Figure 3.1 (adapted from an example in Lawson 2003):

Figure 3.1. Using Toulmin's Model to Support a Claim

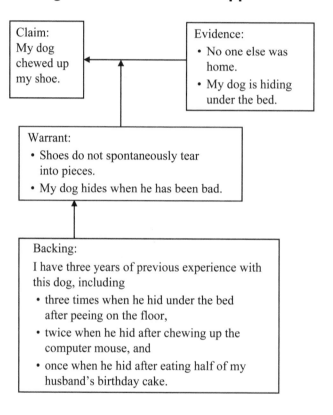

Argument in the Classroom

For the purposes of the middle school classroom, the most important aspect of Toulmin's model is coordinating claims and evidence. Students are surprisingly inexperienced with identifying appropriate evidence to support a scientific claim (Kuhn 1993; Kuhn, Amsel, and O'Loughlin 1998; Zohar 1998). When asked to give evidence to support a claim, students tend to offer reasons their claim is plausible rather than evidence from data. For example, a student group may have done an experiment in which they let cockroaches choose between light and dark environments. Students may make the claim that cockroaches prefer dark environments. They may even have excellent evidence to support that claim, such as that 9 out of their 10 test subjects moved to a dark environment when given a choice. When asked to provide evidence for the claim, however, students may state something like this:

Because cockroaches have those long antennae, maybe they can just feel their way around and don't need light to see.

Now that is a fine hypothesis, and one that could be tested further, but it is not what scientists would use as evidence to support the claim that cockroaches prefer dark environments. Scientists would provide the data that 9 out of 10 cockroaches chose the dark environment. By helping our students understand what counts as evidence for a scientific claim, we help them understand the nature of science.

In science education, we usually use the term *reasoning* to describe the warrant and backing with students; indeed, these are the terms used in the *NGSS* (NGSS Lead States 2013) and the *CCSS* (NGAC and CCSSO 2010). However, understanding the concepts behind the term reasoning can be useful for you as a teacher. For example, as your students become more comfortable with argumentation, they will begin to challenge their classmates' claims and evidence, often on the grounds that the evidence does not adequately support the claim. You will know that they are challenging the warrant, and drawing a chart similar to Figure 3.1 (without the terms *warrant* and *backing*) may help students communicate their critiques more effectively.

Assessing Claims and Evidence

Many of the activities in this book ask students to make a claim and support it with evidence. The rubric in Table 3.1 (p. 28) can be used to assess the claims that students make in these activities and can also serve as a springboard for a class discussion about what makes an effective claim. As students grasp the basic concepts, you might consider moving to a more sophisticated tool to help them think through their claims. The argumentation and evaluation guide, developed by Bulgren and Ellis (2015), is an excellent tool for deepening students' thinking and can serve as a more detailed rubric for assessment.

FIND OUT MORE

For an excellent manual on helping students with claims, evidence, and reasoning, see
- McNeill, K., and J. Krajcik. 2012. *Supporting grade five–eight students in constructing explanations in science.* Boston: Pearson.

For a short and readable introduction, try
- Llewellyn, D., and H. Rajesh. 2011. Fostering argumentation skills: Doing what real scientists really do. *Science Scope* 35 (1): 22–28.

TEACHING TIP

Any simple activity that has students focus on making a claim and supporting it with evidence can be used as a starting point for introducing claims, evidence, and reasoning. For a series of such activities, consider
- Llewellyn, D. 2015. Scaffolding students toward argumentation: Part 1. *Science Scope* 39 (3): 76–81.
- Llewellyn, D. 2015. Scaffolding students toward argumentation: Part 2. *Science Scope* 39 (4): 58–61.

Table 3.1. Rubric for Assessing Claims, Evidence, and Reasoning

Question	Completely	Partially	Not at All
Is it the type of claim that an experiment could verify?			
Does the claim address the question asked in the investigation?			
Does the evidence support the claim?			
Is the evidence sufficient to support the claim, or is more information needed?			
Is the claim correct based on the data from the investigation?			

References

Bulgren, J., and J. Ellis. 2015. The argumentation and evaluation guide: Encouraging *NGSS*-based critical thinking. *Science Scope* 39 (3): 78–85.

Kuhn, D. 1993. Science as argument: Implications for teaching and learning scientific thinking. *Science Education* 73 (3): 319–337.

Kuhn, D., E. Amsel, and M. O'Loughlin. 1988. *The development of scientific thinking skills.* San Diego: Harcourt Brace Jovanovich.

Lawson, A. E. 2003. The nature and development of hypothetico-predictive argumentation with implications for science teaching. *International Journal of Science Education* 25 (11): 1387–1408.

National Governors Association Center for Best Practices and Council of Chief State School Officers (NGAC and CCSSO). 2010. *Common core state standards.* Washington, DC: NGAC and CCSSO.

National Research Council (NRC). 2012. *A framework for K–12 science education: Practices, crosscutting concepts, and core ideas.* Washington, DC: National Academies Press.

NGSS Lead States. 2013. *Next Generation Science Standards: For states, by states.* Washington, DC: National Academies Press. *www.nextgenscience.org/next-generation-science-standards.*

Norton-Meier, L., B. Hand, L. Hockenberry, and K. Wise. 2008. *Questions, claims, and evidence: The important place of argument in children's science writing.* Arlington, VA: NSTA Press.

Toulmin, S. E. 1958. *The uses of argument.* Cambridge: Cambridge University Press.

Zohar, A. 1998. Result or conclusion: Students' differentiation between experimental results and conclusions. *Journal of Biological Education* 32 (1): 53–59.

Reconstructing the Past

Topics

- Making claims from evidence
- Earth science methods
- Dinosaur trackways and behavior

Reading Strategies

- Comprehension coding
- Reading in groups

Lesson Objectives: Connecting to National Standards

The following list shows the *Next Generation Science Standards* (*NGSS*) and *Common Core State Standards* (*CCSS*) supported by this activity.

NGSS: *Science and Engineering Practices*
- Planning and Carrying Out Investigations
- Analyzing and Interpreting Data
- Engaging in Argument From Evidence

NGSS: *Disciplinary Core Idea*
- **ESS1.C.** The History of Planet Earth

NGSS: *Crosscutting Concepts*
- Patterns
- Stability and Change

CCSS: *Literacy in Science and Technical Subjects*
- **CCSS.ELA-Literacy.RST.6-8.1.** Cite specific textual evidence to support analysis of science and technical texts.
- **CCSS.ELA-Literacy.RST.6-8.2.** Determine the central ideas or conclusions of a text; provide an accurate summary of the text distinct from prior knowledge or opinions.
- **CCSS.ELA-Literacy.RST.6-8.5.** Analyze the structure an author uses to organize a text, including how the major sections contribute to the whole and to an understanding of the topic.
- **CCSS.ELA-Literacy.RST.6-8.6.** Analyze the author's purpose in providing an explanation, describing a procedure, or discussing an experiment in a text.
- **CCSS.ELA-Literacy.WHST.6-8.1.** Write arguments focused on discipline-specific content.

Background

This chapter has two main goals. The first is to ease students into the reading procedures described in Chapter 1. For that reason, this chapter has two reading passages. "Chip Bandit" allows students to practice using codes to record what they are thinking as they read. "Lumpheads" can be used to let students practice the group reading process. If you choose not to have your students work in groups, using two reading passages with this lesson will reinforce that reading is an important part of your science class.

The second goal is to introduce students to the idea that one research technique in Earth science is to learn about the past by looking at how things work in the present. Research into the past is often like detective work. Scientists gather evidence and compare it with known information to make a claim about what happened. This kind of research is dependent on the assumption, held across science disciplines, that the laws in the natural world are the same now as they were in the past.

Materials

- Rulers
- Metersticks
- Masking tape
- Chalk or sponges and butcher paper (optional)

Student Pages

- Tracks by the River (lab handout)
- "Chip Bandit" (article)
- On the Tracks of a Dinosaur (lab sheet)
- What to Do in Your Reading Group (strategy introduction)
- "The Lumpheads" (article)
- Looking at Assumptions (thinking visually)

Exploration/Pre-Reading

Begin by showing the class images of dinosaur trackways. These are readily available in a web search. Explain that most trackways formed when dinosaurs left prints along the muddy banks of rivers and lakes, and then sand settled into the footprints and hardened.

Tell students that they have been called in to interpret a new dinosaur trackway that has been found along a river in your state. The footprints are thought to come from medium-sized theropods. Theropods are a group of dinosaurs that include Tyrannosaurus rex. These dinosaurs were bipedal, so they walked on just two feet, like humans and birds. Place students into groups of three and provide each group with a copy of the trackway and a ruler. Give them three minutes to make a list of observations about the trackway.

Allow groups to share some of their observations. It should soon become obvious that the distance between footprints gets bigger toward the end of the track. Tell students that the distance between footprints is called stride length. Guide students to formulate a question similar to the following: What could cause the stride length to increase in this trackway? Have students propose possible explanations. Explanations could include that the surface changed in some way, that three different dinosaurs left the tracks, or that the dinosaur changed speed. Get students to focus on the hypothesis that the dinosaur may have changed speed. Point out that this is a hypothesis that you have the means to explore further.

Introduce the First Reading. Tell students that you will come back to testing this hypothesis after they have read a story about a way that Earth scientists figure out what happened in the past. Introduce the idea of using codes as described in the Reading Strategies section. Then have students read "Chip Bandit."

Reading Strategy 1 (for "Chip Bandit"): Comprehension Coding

To introduce the strategy, use a projector to display a copy of "Chip Bandit" so students can watch as you code. Read the first four paragraphs out loud and model the coding process. For example, you might read the first four sentences and put a question mark, saying, "This is confusing to me because I was expecting to read about dinosaurs, not dogs." At the mention of the second dog, you could put an exclamation point and say, "Oh, here's the mystery. I bet this is going to be important later on." Finally, at the mention of Earth scientists, you can put another exclamation mark, saying, "I see. Here's the connection to Earth science. I bet this is important." Point out that you did not put a code next to every sentence; students only need to place codes where they seem appropriate.

After modeling the strategy for a few paragraphs, give students time to read and code the rest of the article independently.

Further Exploration/Pre-Reading for Second Article. After students have read the article, hand out "On the Tracks of a Dinosaur." Review the question and the hypothesis that students are investigating. Ask them to consider what kind of data they could collect to see if speed could affect stride length. If they have trouble coming up with responses, ask what data Tyrone and Zachary used in the story to learn more about the dog footprints and what data the scientists used to figure out how dinosaurs walked. Someone should soon suggest looking at how speed affects stride length in modern animals. Point out that you happen to have a classroom full of modern, middle school, bipedal animals.

Either in small groups or as a class, have students devise a method for measuring stride length. Students could walk a set distance, count how many steps they take, and then divide the distance by the number of steps. They could take a set number of steps and measure the distance needed. They could tape sidewalk chalk to their shoes and measure the distance between marks. They could even tape wet sponges to their feet and walk on butcher paper to produce footprints to measure.

Once they have designed a system, have them draw a diagram of their plan. I find that students carry out their procedure more accurately if they have drawn a picture rather than just describing their plan in words. Then have them make a data table and collect data. Once each group has their measurements, compile data as a class. Since measuring systems may vary between groups, the simplest way to compile the data is to get a count of how many people had a longer running stride than walking stride.

At this point, students should be ready to make a claim and support it with evidence from their experiment and the class data. If students ask, you can also tell them that other studies have shown that a variety of animals, including camels, dogs, ostriches, sheep, and elephants also have longer strides when they are running.

Introduce the Second Reading. Tell students they are going to read a second passage that describes research on another dinosaur mystery. As they read, they should look for how the scientists used things from the present to make sense of the past. Show them the word *pachycephalosaurid*, pronounce it for them, and explain that it is a type of dinosaur. For this first discussion, you don't want the students to feel awkward about pronunciation, so tell them that they can call it the pack dinosaur if they wish.

> **SAFETY NOTE**
> If students use wet sponges, use caution not to create a slip or fall hazard with the water on the paper.

> **TEACHING TIP**
> Watch out for claims that attempt to say why the dinosaur was running, such as, "It saw a predator and ran." Point out that they do not have evidence to support those types of claims.

Reading Strategy 2 (for "Lumpheads"): Group Reading Procedure

To introduce this strategy, hand out What to Do in Your Reading Group to each student. Read over the procedure with your class, then select three confident students to come to the front of the class and model the procedure for the first section of text (down to the first solid line). Then place the rest of your students into groups of three to try the process themselves. Have them start from the beginning of the article to reinforce the process they just watched.

Remember that the first time students use a procedure like this, it may not go smoothly. Students will benefit most from the process after they have had a chance to practice it two or three times.

FIND OUT MORE

To learn more about what we can discover about dinosaur movement from trackways and fossils, try *Dynamics of Dinosaurs and Other Extinct Giants* by R. McNeil Alexander. It is an older book, but most of it is still valid, and it is presented well for laymen. For more on Joseph Peterson's research, which is referenced in "Lumpheads," you can view his journal article online.

- Alexander, R. M. 1989. *Dynamics of dinosaurs and other extinct giants.* New York: Columbia University Press.
- Peterson, J. E., C. Dischler, and N. R. Longrich. 2013. Distributions of cranial pathologies provide evidence for head-butting in dome-headed dinosaurs (pachycephalosauridae). *PLoS ONE* 8 (7): e68620. *http://journals.plos.org/plosone/article?id=10.1371/journal.pone.0068620.*

Journal Question

What did you do today that helped your reading group? Did you do anything that was unhelpful? What could you do next time to help your group even more?

Application/Post-Reading

- Thinking Visually: Looking at Assumptions
- Writing Prompt: The writing assignment for this chapter is the claims and evidence portion of the lab.
  - Key Evaluation Point: The most likely claim is that the dinosaur stride length increased because the dinosaur began to move more quickly. However, if student evidence supports another claim, it may be valid.

Tracks by the River

Chip Bandit

Tyson and Zachary grabbed some snacks and headed out for the basketball court.

"I'm going to bring Calvin," said Tyson. Calvin was Tyson's new dog. He was an energetic Collie, and was always getting into trouble if Tyson left him at home.

"No problem," said Zachary, "as long as she stays out of our snacks. I've been looking forward to these chips."

As they got to the park, a little mutt ran up and greeted Calvin. The boys left the dogs to play together and hit the court. When they returned to the table where they had put their things, the potato chip bag was shredded, and there wasn't a single chip left.

Two dogs playing in the park

"Calvin ate the chips!" yelled Zachary. "You are so buying me another bag."

"No way!" answered Tyson. "He would never tear into an unopened bag of potato chips. I bet it was that other dog, the little mutt."

The Problem

Tyson and Zachary had a problem. They didn't know which dog ate the chips, and they couldn't go back in time to watch! They would need to do some detective work.

Earth scientists face the same kind of problem when they are piecing together the history of the Earth. Most of Earth's history took place long before any people were around to see it. So Earth scientists collect clues to the past and then use them to figure out what must have happened.

"Look," said Zachary. "There are footprints in the dirt all around the table. Dog footprints. *Big* dog footprints. That other dog was little. I think Calvin ate the chips."

"There are little dog footprints mixed in here, too," said Tyson. "I still say it was the mutt."

"Well, one of them climbed up on this side of the table, grabbed the bag, and dragged it to the ground. I think we just need to figure out which prints are headed toward the table," said Zachary. "But how can we tell which direction the dogs were walking?"

Tyson thought for a moment. Then he slid off his own shoes and walked through a patch of bare dirt. "Look at my footprints," he said. "The toes always point in the direction that I'm going." He walked back over to the prints near the table. "The dog prints have toes, too. And the toes on these little paw prints are headed straight for the table—right to where the chips were."

NATIONAL SCIENCE TEACHERS ASSOCIATION

Problem Solving

Tyson figured out which way the dogs were walking by comparing the dog prints to his own footprints. Earth scientists do something similar. They use what they know about how the world around them works to make sense of clues from the past.

For example, paleontologists, or Earth scientists who study ancient life, wondered about how dinosaurs walked. Did dinosaurs walk with their legs spread out to the sides of their body like a modern lizard? Or did they have their feet right under their body, more like a bird or mammal?

Paleontologists had two types of evidence to consider. They had the fossilized leg and hip bones of a variety of dinosaurs. They also had lines of dinosaur footprints, called trackways. Dinosaur trackways formed when dinosaurs left prints along the muddy banks of rivers and lakes and then sand settled into the footprints and hardened.

Paleontologists looked at the hips and legs of the dinosaurs and compared them to both lizards and birds. The dinosaur bones were shaped more like bird hips and legs than lizard hips and legs. Scientists compared the dinosaur trackways to tracks made by modern lizards and birds (Figure S4.1). Lizard tracks are spaced wide, reaching out from either side of the body. Dinosaur footprints are located close together, like a bird's. The paleontologists concluded that dinosaurs walked with their legs directly underneath their bodies.

Chip Time

Back at the park, Zachary shook his head and looked at Calvin. "Well, dog, I guess the evidence is on your side. Come on, I'll pick up another bag of chips on the way home. Maybe I'll even share them with you."

Figure S4.1. Footprint Patterns for Different Animals

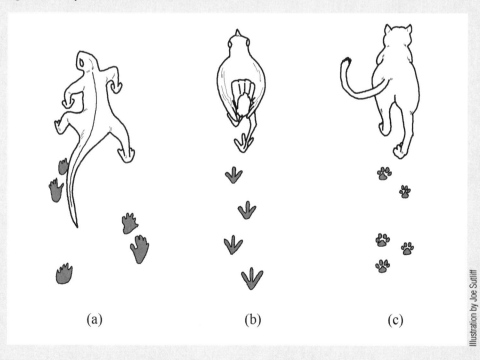

(a)   (b)   (c)

Illustration by Joe Sutliff

On the Tracks of a Dinosaur

You have been called in to interpret a new dinosaur trackway that has been found along a river in your state. The footprints are thought to come from medium-sized theropods. Theropods are a group of dinosaurs that include Tyrannosaurus rex. These dinosaurs were bipedal, so they walked on just two feet, like humans and birds.

Background Information
Record your most important observations of the dinosaur trackway here.

Question
What caused the stride length to increase along the trackway?

Hypothesis Under Consideration
The stride length increased because the dinosaur changed speed.

Procedure
Draw and label a diagram of how you will measure stride length.

You will need to measure the stride length walking and running for each member of your group. *What things will you need to keep the same each time to make it a fair test?*

Results
Make a table to fill in as you collect data with your group.

Class Data

Summarize the findings for all the groups in your class.

Other Data

Summarize the information from the graph your teacher shows.

Conclusions

Make a claim: What caused the stride length to increase in the dinosaur trackway?

What evidence supports your claim?

How does that evidence support your claim?

What to Do in Your Reading Groups

To begin, pick a leader, an emergency manager, and an interpreter.

1. Everyone reads the first section quietly and marks !, ✓, ×, and ? while reading.
2. The *leader* asks each member of the group to share anything that was confusing (marked ? or ×).
3. The group should try to figure out and explain what the confusing word, sentence, or idea means.
4. If the group cannot make sense of the confusing word, sentence, or idea, the emergency manager should raise the flag to get help from the teacher. Keep working until the teacher is available.
5. Repeat steps 1–4 for Section 2 (separated by a line).
6. Repeat steps 1–4 for Section 3 (separated by a line).
7. The group should work together to answer the Big Question. The *interpreter* will write the group's answer to turn in to the teacher.

The Lumpheads

Pachycephalosaurids were the lump-heads of the dinosaur world. Their skulls could reach two feet long, but they housed a marble-sized brain. The main feature of their heads was a dome of solid bone that bulged from the top.

Naturally, scientists have wondered what the lump was for. One hypothesis is that the lump helped attract a mate. In effect, a big, impressive lump would be saying, "Hey, look! I've got so much extra food and energy to spare that I can waste it on this gorgeous lump. Just imagine what good genes I can pass on to our babies." Another hypothesis holds that the lump was a weapon. In this hypothesis, dinosaurs used the lump to smash into rivals to chase them away from good territory and potential mates.

Pachycephalosaurid skull

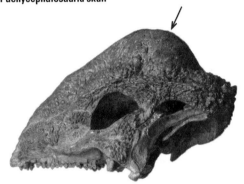

Note: The arrow points to the lump on this pachycephalosaurid skull.

Unfortunately, no one was around to see what pachycephalosaurids did, so there is no way to observe the lumps in action.

Pits in the Skulls

Enter Joseph Peterson. Peterson's first job was at a museum where he participated in a dinosaur dig in Montana. The team discovered a gigantic pachycephalosaurid skull, and Peterson noticed that it had two large pits on top of its dome. Several years later, he and a team of researchers revisited the skull to see if the indentations could help them understand how the dome was used.

The pits reminded Peterson of bone damage he had seen on animals that had severe infections. When a modern animal, such as a cow, has a bone injury, it can leave a pit. When the pit begins to heal, the bone created in the healing process is less dense than the original bone. Peterson assumed the same would be true for dinosaur bone.

He collected more than 100 pachycephalosaurid skulls and took some to a hospital for a computerized tomography (CT) scan. Sure enough, the pits had areas of less-dense bone that had been growing

Artist's representation of an injured pachycephalosaurid

Artist's representation of fighting pachycephalosaurids

back over the damaged spots. Peterson and his team concluded that the pits came from injuries that started healing while the dinosaurs were still alive.

What Kind of Injuries?

The next question was whether those injuries were related to the function of the dome, or if they were just accidents. They found that almost 25% of the skulls of domed pachycephalosaurids that they examined had evidence of injury. However, some pachycephalosaurids did not have domed skulls. These were probably female or young dinosaurs. None of the pachycephalosaurid skulls without domes had injuries. Clearly, the injuries to the domed skulls were not random accidents. So what were the dinosaurs doing with those lumps that made them so prone to injury?

Once again, the researchers looked for clues in modern animals. Modern animals that fight with their heads do so in three ways. Goats ram their heads into the sides of other goats. Cows and American bison lock horns and wrestle, trying to knock their opponent over. Dall and bighorn sheep slam their heads together when they fight. The researchers gathered sample skeletons from these types of animals and examined them for pits that would indicate bone injury.

Modern Damage Patterns

As the scientists expected, most of the injuries to goats were found along their spine and ribs. The cows and bison had a few head injuries but more side injuries from being knocked over. The Dall and bighorn sheep had the most damage to their skulls. Additionally, the shape of those injuries were most similar to the shapes of the injuries found on the pachycephalosaurids. Peterson and his team concluded that the domes were used for fighting in head-to-head ramming contests.

Throughout their investigation, the researchers relied on a common tool used in Earth science to uncover the past. They used evidence from modern animals and processes that we *can* observe to make sense of events in the past that we cannot observe.

And as for the idea that the domes might have been used as a display to attract mates—Peterson points out that it could still be true. Certainly, carrying a large weapon on your head sends a clear a message about your strength and power to potential mates. For a small-brained dinosaur, that might not be such a lumpheaded move after all.

THE BIG QUESTION

What modern comparisons did the researchers use to answer their questions?

NATIONAL SCIENCE TEACHERS ASSOCIATION

Looking at Assumptions

In this research, Peterson used modern day animals as models for what might have happened with dinosaurs. All models make assumptions. Fill in the outline of his research below and pay special attention to how his assumptions are important to his claim.

How did pachycephalosaurids use those domes?

Hypothesis (Possible Claim) #1: Domes helped attract a _____.
Hypothesis (Possible Claim) #2: Domes were used for _____.

Peterson found a clue! Many of the skulls had pits on their domes. He analyzed the clue in two ways:

- In modern animals, bone that is healing is different from other bone because it is _____.

- What did Peterson find when he looked at the bone density in the pits on the dinosaur skulls?

- How did that support the claim that the pits came from fighting?

- **The Assumption:** Bone injuries in _____ and _____ heal similarly.

- Some modern animals use their heads to fight. Each style leaves a different pattern of _____.

- What did Peterson find when he compared patterns on the dinosaurs with those on the skulls of modern animals?

- How did that support the claim that the pits came from fighting?

- **The Assumption:** Fighting in _____ and _____ leave similar patterns of injuries.

Do you think these assumptions are reasonable? Why or why not?

How might Peterson change his claim if future research disproves one or more of the assumptions?

Mountain Mayhem

Topics
- Erosion and deposition

Reading Strategy
- Finding the meaning of new words

Lesson Objectives: Connecting to National Standards

The following list shows the *Next Generation Science Standards* (*NGSS*) and *Common Core State Standards* (*CCSS*) supported by this activity.

NGSS: *Science and Engineering Practices*
- Developing and Using Models
- Constructing Explanations and Designing Solutions

NGSS: *Disciplinary Core Idea*
- **ESS2.C.** The Roles of Water in Earth's Surface Processes

NGSS: *Crosscutting Concepts*
- Cause and Effect
- Stability and Change

CCSS: *Literacy in Science and Technical Subjects*
- **CCSS.ELA-Literacy.RST.6-8.4.** Determine the meaning of symbols, key terms, and other domain-specific words and phrases as they are used in a specific scientific or technical context relevant to grades 6–8 texts and topics.
- **CCSS.ELA-Literacy.RST.6-8.9.** Compare and contrast the information gained from experiments, simulations, video, or multimedia sources with that gained from reading a text on the same topic.
- **CCSS.ELA-Literacy.WHST.6-8.2.** Write informative/explanatory texts, including the narration of historical events, scientific procedures/experiments, or technical processes.

Background

This lesson is designed to introduce students to the key ideas and vocabulary associated with erosion. Students will see erosion at work in a simulation and try out some different ways to lessen the effect. Then they will read about the three main agents of erosion. Students often see erosion as an entirely destructive force, so one goal of this lesson is to help them see that the processes of erosion and deposition lead to both breaking down and building up, and are, in themselves, neither good nor bad. The process of weathering is not addressed in this lesson, so you may want to consider that topic before or after using these materials.

Materials

- Bag of play sand
- Large bucket
- Trays or small tubs (1 per group)
- Cups of water (1 per group)
- Drinking straws (1–2 per group)
- Spray bottles of water (1 per group)
- Ice cubes (1 per group)
- Popsicle sticks (10 per group)

- Tape (1 roll per group)
- Sanitized indirectly vented chemical-splash goggles

Student Pages

- Mountain Mayhem (lab sheet)
- "The Water That Hauled Off a Highway" (article)
- On the Move (graphic organizer)

Exploration/Pre-Reading

Before class, prepare a "mountain" for each group by piling damp sand in the tray or tub. Use your fingers to smooth a flat place to serve as road through the middle of the mountain, as shown in Figure 5.1.

Figure 5.1. Mountain Model Setup

> **TEACHING TIP**
> If you wish to reuse sand between classes, have the students dump their sand into the large bucket. Most of the water will drain to the bottom of the bucket in 15–20 minutes. Then you can scoop sand back into each group's container. You may also wish to have a rag for each group to wipe off their table when they are done and a thank-you note for your custodian that evening.

Tell students that today is their chance to be a destructive force. Give each group a tray with a "mountain road," a cup of water, a plastic straw, a spray bottle, and an ice cube. Instruct them to cause as much damage to their road as they can, while abiding by the rules listed on their lab sheet.

As they use each tool, they should fill in their Mountain Mayhem data sheet. Then, they should complete the mini-engineering task to try to build a structure to protect the road.

Introduce the Reading. Tell students that they are about to read an article that will talk about how real mountains can break down.

Reading Strategy: Finding the Meaning of New Words

To introduce this strategy, ask students if they have ever come across a new word in a science book. Tell them that it can be hard to read science books because of all of the new words, but the nice thing about science writing is that the text usually tells you what the new words mean. You just have to know how to recognize the definitions.

Display Table 5.1 for students. Have students read through the table. Then ask if they can think of other ways a definition might be given. Add any new ideas to the chart. Point out that definitions are usually given just before or after a new word is used for the first time. Have student watch for new words and their definitions as they read. In this article, the three key words that will be introduced are erosion, sediment, and deposition. Delta, terraces, and glaciers are also defined using signal words from Table 5.1.

Table 5.1. Common Ways That Texts Introduce New Words

Example	Explanation
Soil can be washed away by runoff. Runoff is the water that collects and moves across the ground during a rainstorm.	The sentence after the term provides a definition.
Water that moves across the ground during a rainstorm, called runoff, …	The new term is signaled with the word *called*.
Soil can be washed away by runoff, which is rainwater that runs across the ground.	The definition is signaled with the phrase *which means* or *which is*.
Runoff, or water that runs across the ground in a rainstorm, …	The word *or* after a comma indicates that the word and phrase mean the same thing.
Soil can be washed away by runoff. This movement of rainwater across the ground …	This is the trickiest situation. The text doesn't directly say what the word means, but implies it by using the word and the definition close together.

Journal Question

In today's reading, you looked for ways that definitions can be given for new words. Now you try writing a paragraph that includes a definition. Think of a science word that a younger child might not know. Using one of the examples from Table 5.1 as a pattern, write a paragraph that introduces the meaning of your word.

Application/Post-Reading

- Graphic Organizer: On the Move
- Writing Prompt: Digger Johnson has just gotten a contract to build a road into the side of a mountain. He doesn't know a thing about erosion. Write Mr. Johnson a letter explaining how erosion could affect his road. Give him some suggestions for protecting it.
  - Prewriting Questions: Remember that this contractor does not know about erosion. What are some things you'll want to be sure to tell him? What science words should you include? What kind of language should you use (everyday language or school/business language)?
  - Key Evaluation Point: Students should describe how erosion is the process of moving soil and rocks to a new place. It can be slowed by planting bushes or grasses, by building terraces, or by installing walls and fences along the mountainside.

Mountain Mayhem

Become a destructive force! Cause as much damage to your road as you can, following the guidelines below:

1. Use only the tools you are given.
2. Do not touch the sand, container, or table.
3. The only tool that can touch the sand directly is your ice cube. You can set it in one place and give it one flick with your finger.

Tool	What You Did With It	What Happened
Squirt Bottle		
Cup of Water		
Straw		
Ice Cube		

Mini-Engineering Challenge

Rebuild your mountain and road. Using the Popsicle sticks and tape, design a structure to protect the road from damage by the squirt bottle. Sketch it below.

Use the squirt bottle to test your design. How well did it protect your road?

Now that you've seen your design in action, what changes do you think might make it more effective?

The Water That Hauled Off a Highway

On August 28, 2011, Tropical Storm Irene swept into Vermont, dumping rain. Almost every river and stream in the state overflowed its banks. Entire sections of Route 4 washed away, leaving cliffs as tall as a two-story building in their place. The people in the town of Killington were trapped, with no roads left for driving in or out.

Destruction caused by Hurricane Irene

Dirt, mud, and rocks on the surface of the Earth are constantly shifting their location. The flooding in Vermont simply sped up a process that sometimes takes hundreds or thousands of years. It's the process of erosion: the movement of rock and soil from one place to another.

How Erosion Happens

Gravity is the driving force of erosion. Consider a landslide in which rocks tumble off a mountain. Which way do the rocks fall? Down! The rocks are pulled down by the same attractive force that pulls everything toward the Earth.

Water is the most common vehicle of erosion. Individual raindrops can scatter dirt. Water collects on parking lots and then races off the side as a sheet of water. Rivers rub against their banks and drag off loose dirt and wear away rock. The dirt and rock that gather in the water are called sediment. Eventually, the water slows down and the sediment sinks and is left behind.

Wind also contributes to erosion. Think of a sand storm in a desert. Wind whips across the ground with such force that sand is pushed into the air and driven in the direction the air is moving. When the wind dies down, the sand falls and leaves behind a sand dune.

Sand dune in Utah's Coral Pink Sand Dunes State Park

Even ice causes erosion. During an ice age, blocks of ice miles wide and high build up from snow that never melts. These glaciers creep across the landscape at a rate of only a few inches a year, pushing debris along in front. Glaciers can scoop out hollows in a mountainside and create lakes along their path.

Slowing the Destruction

While some agents of erosion, such as the movement of glaciers or the force of a tropical storm, are hard to stop, more gradual erosion can be limited. Construction teams often plant bushes or grass along the hillside near a road. The plants slow down the rainwater as it runs by. The roots of the plant loosen the soil so that it is easier for the water to soak in.

Terraces on a mountain

Water can also be slowed down by building a series of steps into the side of a hill. These steps are called terraces, and mountain farmers have used them for centuries to keep their rich topsoil from washing away. Erosion by wind can be slowed by putting obstacles in the way. Many cities have built fences on their beaches to slow the wind and collect sand as it blows by.

Building Something New

People try to stop erosion when it is destroying something of value, but the process of erosion has another side. The dirt and rocks that have eroded get dropped off somewhere new, in a process called deposition. Deposition comes from the word deposit, which means "to leave something behind."

For example, rivers deposit sediment when they meet the ocean. Eventually, enough sediment is deposited in one place to create new land. Land formed from deposition at the edge of the ocean is called a delta. Deltas have rich soil and make excellent farmland. Even the rocks and soil washed from beneath the highway by Tropical Storm Irene were deposited, and, somewhere, created something new.

THE BIG QUESTION

Think about your lab. Describe one thing that happened in the lab using the words *erosion*, *sediment*, and *deposition*.

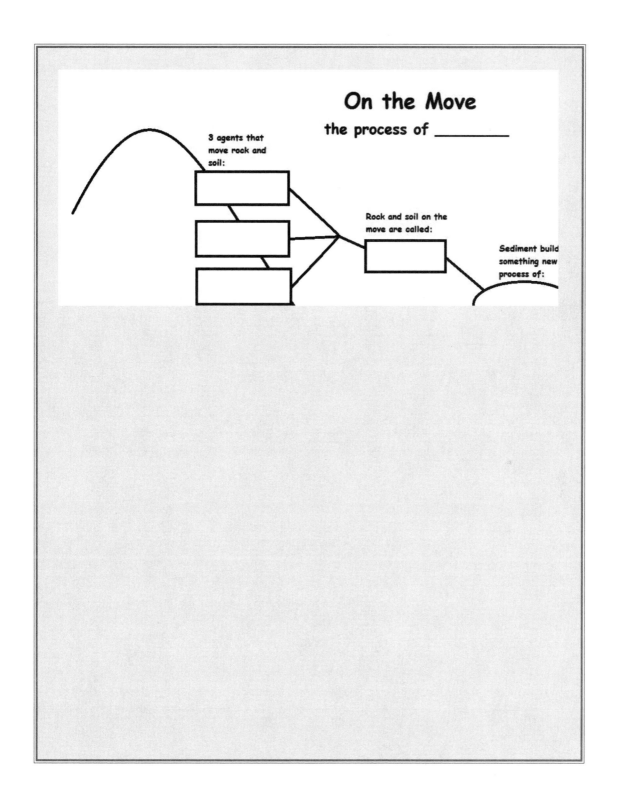

On the Move
the process of _____

3 agents that
move rock and
soil:

Rock and soil on the
move are called:

Sediment build
something new
process of:

Continents on the Move

Topics
- Plate tectonics
- Alfred Wegener
- Nature of science

Reading Strategy
- Chunking

Lesson Objectives: Connecting to National Standards

The following list shows the *Next Generation Science Standards* (*NGSS*) and *Common Core State Standards* (*CCSS*) supported by this activity.

NGSS: *Science and Engineering Practices*
- Analyzing and Interpreting Data
- Engaging in Argument From Evidence

NGSS: *Disciplinary Core Ideas*
- **ESS1.C.** The History of Planet Earth
- **ESS2.B.** Plate Tectonics and Large-Scale System Interactions

NGSS: *Crosscutting Concept*
- Cause and Effect

CCSS: *Literacy in Science and Technical Subjects*
- **CCSS.ELA-Literacy.RST.6-8.2.** Determine the central ideas or conclusions of a text; provide an accurate summary of the text distinct from prior knowledge or opinions.
- **CCSS.ELA-Literacy.RST.6-8.7.** Integrate quantitative or technical information expressed in words in a text with a version of that information expressed visually (e.g., in a flowchart, diagram, model, graph, or table).
- **CCSS.ELA-Literacy.WHST.6-8.1.** Write arguments focused on discipline-specific content.
- **CCSS.ELA-Literacy.WHST.6-8.2.** Write informative/explanatory texts, including the narration of historical events, scientific procedures/experiments, or technical processes.

Background

Plate tectonics is the primary theory that drives explanation in geology, but the idea that land masses drift around the Earth can sound as crazy to students as it did to geologists in the early 1900s. Spending time on the history and evidence of plate movement can help students understand this pivotal idea. In this chapter, students will consider some of Alfred Wegener's evidence for continental drift and read about what it took to challenge the prevailing views in geology.

Materials

- Magazine advertisement (1 per group)
- Set of southern Pangaea continents (1 per group)
- Envelopes or plastic sandwich bags

Student Pages

- Continents on the Move? (lab sheet)
- "Wegener's Bold Claim" (article)
- Seafloor Spreading (thinking visually)

Exploration/Pre-Reading

Before class, cut the magazine advertisements into about six pieces. Remove one piece so that students do not have a complete puzzle. Place each advertisement into an envelope. Also, cut out the southern Pangaea continents and a copy of the key for each group and place these in another envelope. The continents do not have to be cut perfectly along all dips and curves. Both sets can be used across multiple classes.

Begin by having groups try Part 1 of Continents on the Move?, in which they look for evidence that the pieces of a magazine advertisement come from the same page. This will help them think about what kind of evidence would suggest that the continents were once connected. Then have groups complete Part 2, in which they consider some of Wegener's evidence for continental drift.

Introduce the Reading. Tell students that they are going to read more about Alfred Wegener and the ideas he proposed. You may want to show them Greenland on a map and explain that, despite its name, it is a cold, icy island.

Reading Strategy: Chunking

To introduce the strategy, put the following sentence on the board:

Glaciers leave behind rock deposits as they move, and sometimes leave deep scratches in the bedrock.

Point out that this sentence, like many sentences in science writing, has a lot of ideas crammed into a short sentence. It might be difficult to understand all of the ideas at one time, but if students break the sentence into chunks, they can think about each piece individually.

Add slashes (/) to the sentence on the board so it reads like this:

Glaciers / leave behind rock deposits as they move, / and sometimes leave deep scratches in the / bedrock.

Talk them through the sentence, one section at a time. Start with the word glaciers. Ask, "What is a glacier?" Then look at the next section. The phrase "rock deposits" may be difficult. Point out to students that they can visualize an image of glaciers leaving bits of rock behind as they move.

Ask if anyone has questions about the glaciers leaving scratches. Have them visualize scratches in a rock. Would it be easy to scratch a rock? Would scratches be preserved for a long time? What kind of rock are the

> **TEACHING TIP**
> If your students are comfortable with using claims and evidence, introduce this counterclaim in Part 1: *You can't know if the pieces came from one page because you don't have the complete page.* Help them rebut the counterclaim by asserting that their evidence is still valid, even if they do not have all possible information.

scratches in? Make sure students know that bedrock is the solid rock under the dirt and loose rock we usually see.

When you are finished, summarize the information you have gathered from this sentence, saying something such as, "So, these ice masses drop rocks and boulders as they move. They even scratch the hard rock underneath the ground, leaving marks that people could find later."

Explain that chunking a sentence is like eating a pie. People cannot put the whole pie in their mouth at one time; everyone eats it bite by bite. When eating, some people will take bigger bites than others. Some people will need to break a sentence into more chunks than others, and that is okay. For this article, students can separate the chunks using slashes, like you did on the board. When they are reading something they can't write on, they can chunk it in their head or cover up the parts of the sentence they aren't thinking about.

Journal Question

Chunking is especially useful when you are reading long sentences full of new information. Think of a topic you know a lot about. Write a sentence that gives a lot of information on your topic. Use slashes to mark how your reader might chunk that sentence.

> **TEACHING NOTE**
> You may need to explain to students what happens at a scientific conference.

> **FIND OUT MORE**
> To learn more about Wegener's life and theory, see
> • McCoy, R. M. 2006. *Ending in ice: The revolutionary idea and tragic expedition of Alfred Wegener.* Oxford: Oxford University Press.

Application/Post-Reading

- Thinking Visually: Seafloor Spreading
- Writing Prompt: Imagine that you could go back in time and talk to the geologists at the conference where they mocked Wegener's idea. Explain to them how new evidence from the ocean and satellites supports the idea that continents can move.
  - Prewriting Questions: Jot down the types of evidence you want to mention in your speech. Think of an opening sentence that would introduce your ideas and a closing sentence to summarize your points. What science words will you want to include? What are some writing words you might use? (*therefore, in conclusion*)
  - Key Evaluation Point: The mountains in the ocean are made of young rocks where magma is seeping through gaps between the plates. Satellites can measure the movement of land on Earth.

Continents on the Move?

Part 1: Look at the pieces of paper provided by your teacher. Do you think they were ever part of the same page?

Claim (circle one): The pieces of paper (were / were not) originally part of the same page.

Evidence:

How does this evidence support your claim?

Part 2: In the early 1900s, a man named Alfred Wegener studied a variety of geological puzzles around the Earth. He wondered if they might all be clues to the past. Three of his puzzles are described below.

A. Mountain Ranges

The Appalachian Mountains are a very old mountain range that runs along the eastern United States. Mountains that are very similar in age and formation run through the British Isles and Northern Europe. There are no mountains in the ocean between them. Wegener wondered what would have crumpled the land into mountains in two places while leaving the ground beneath the ocean untouched. Look at the two maps in Figure S6.1

Figure S6.1. Mountains Running Through the British Isles and Northern Europe

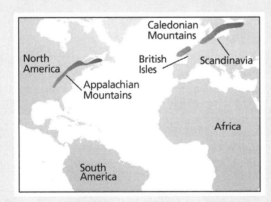 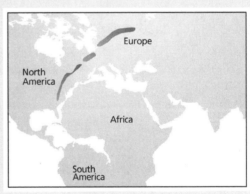

How does combining the continents solve the problem of how the mountain range could have formed?

B. Glaciation

As glaciers move across land, they leave scars and scratches in the rock that show the direction they are moving. Wegener was puzzled by the presence of glacial scarring in places that seemed too close to the equator to have ever been cold enough for glaciers. He also noticed that the scars all showed movement in the same general direction. Look at the map in Figure S6.2. Dark areas show where there is evidence of glaciers, and arrows show the direction of movement.

Figure S6.2. Glacier Movement

How could you arrange the continents to solve Wegener's two puzzles: that the areas with glaciers are far away from each other and that some of the areas with glaciers are very near the equator?

C. Fossils

A third puzzle had to do with the locations of fossils that dated back to about 225 million years ago. Take *Glossopteris.* It was a plant: it couldn't pick itself up and walk to new places. Its spores were fragile and wouldn't survive a long trip. Birds hadn't evolved yet, so there were no animals to carry the plant across the ocean. Yet collections of *Glossopteris* fossils are located on all of the southern continents. The same was true of several other fossils from that time period. Use the cutouts of the locations where fossils were found to find a possible arrangement of the land around 225 million years ago.

Draw or trace your shapes to create a diagram of the arrangement you come up with.

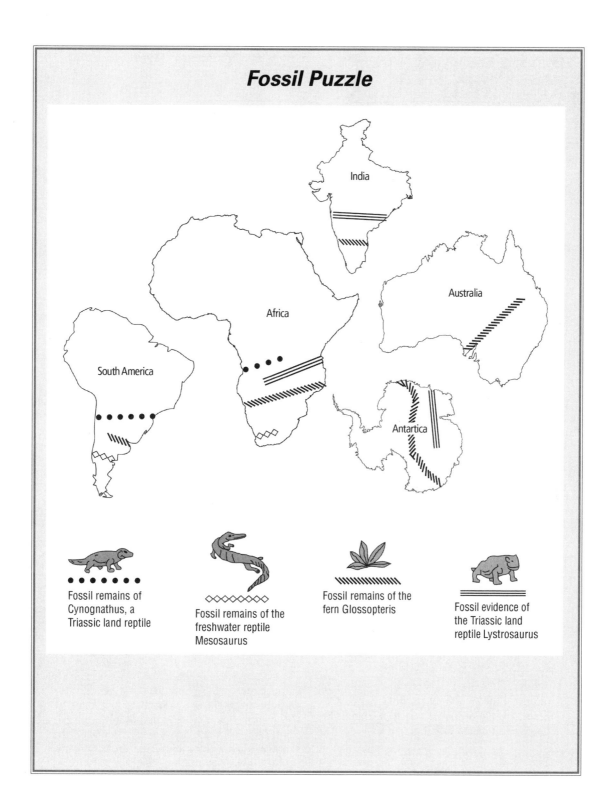

Fossil Puzzle

Fossil remains of Cynognathus, a Triassic land reptile

Fossil remains of the freshwater reptile Mesosaurus

Fossil remains of the fern Glossopteris

Fossil evidence of the Triassic land reptile Lystrosaurus

Wegener's Bold Claim

Alfred Wegener was puzzled. He studied climates—ancient climates, to be exact. One bit of information had bothered him for a long time. The parts of the world that had coal were not where they should have been. Coal forms in tropical areas where dead ferns and other warm-weather plants have been compressed into rock over millions of years. But Wegener knew that coal deposits were found in places too close to the Arctic to have ever been warm enough for tropical plants. There were even coal deposits beneath the frozen tundra in Siberia.

Glaciers bothered him, too. At about the same time that coal was forming in places it didn't belong, there seemed to have been glaciers in places they didn't belong. Glaciers leave behind rock deposits as they move and sometimes leave deep scratches in the bedrock. Wegener saw evidence of glaciers in places that were too hot for ice.

As he read the work of other scientists, he learned about other puzzles. The Appalachian mountain range started in North America, seemed to disappear at the edge of the ocean, and start up again in Europe. Identical fossils of plants and animals that could not swim across the ocean were spread across multiple continents. These fossils all dated from about the same time period. Later fossils from those areas were quite different.

Wegener suspected the continents had moved. But how could whole continents move? If he proposed such a thing, he would risk looking crazy.

Pioneer or Daredevil?

Wegener was no stranger to risk. Even as he thought about his puzzles, he was off on daring explorations. At age 26, he and his brother set a hot air balloon record, staying aloft for more than two days. That sounds minor in today's world of easy air travel, but back then it meant hanging loosely above the Earth in a fragile wooden basket—with only a rough ability to navigate—and hoping that your balloon did not tear or break and send you plunging back to Earth.

Wegener worked on his continental drift theory over the long winter of 1913. He and his research partner were the first explorers to spend the whole winter in the center of Greenland. They wore thick coats (Figure S6.3). They built themselves a shelter out of plywood and packed it in snow for insulation. It was only about the size of a two-car garage, and they shared it with the five ponies that had hauled in their building materials. Survival was not guaranteed. On his previous expedition to Greenland, three of Wegener's colleagues had frozen to death on the trip home.

Figure S6.3. Thick Coat Worn by Wegener in Greenland

When he got home from Greenland, he got ready to share his ideas on continental drift. It was a good thing Wegener was tough.

Challenging the Hot Potato

He wasn't the first scientist to wonder about the continents. Others had noted that South America and Africa look an awful lot like they should fit together. But Wegener was the first to compile evidence from several science areas and put forth the claim that, in fact, the continents had moved. In 1915, he published *The Origin of Continents and Oceans.* Among other things, it proposed that mountains were formed by continents crashing into each other and folding the Earth up.

Geologists were not impressed. Who was this climate scientist claiming to tell *them* about the history of the Earth? It's not that they had everything figured out. The current theory in geology was that the crust of the Earth was cooling and mountains formed because the land was wrinkling like the skin of a baked potato. Geologists knew their theory had a problem, because mountains should be everywhere instead of mostly at the edges of continents. But they were not going to let an outsider change their views. They also had a valid criticism of Wegener's idea. He couldn't figure out what force would be strong enough to propel the continents across the ocean floor. At a conference held to discuss continental drift, geologist after geologist took the floor to ridicule Wegener. For years, Wegener's idea was a joke among geologists, and comparing someone to Wegener was considered an insult.

Sea Floor Spreading

His evidence wasn't going away, however. And after World War II, a new piece of the puzzle was found. When governments were mapping the ocean floor in order to steer their submarines, they found enormous mountains that stretched across the middle of each ocean like the stitching on a baseball (Figure S6.4). The rocks that made up these mountains were some of the youngest on Earth. It appeared that the floor of the ocean was pulling apart, and magma from deep in the Earth was oozing up to form mountains. Suddenly Wegener's idea didn't seem so crazy. Only, it wasn't just continents moving around the Earth and having to plow through the ocean. The whole crust of the Earth was broken into plates that could slide, ever so slowly, into new places. As technology has improved, geologists have even confirmed plate movement using measurements from satellites in space.

Figure S6.4 Locations of Mountains Across the Ocean Floor

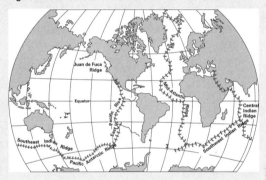

Wegener did not live to see his theory accepted. He continued researching and rewriting his book for the rest of his life. But he also poured himself into climate research in Greenland, and his work there was well accepted. In fact, he died there on his fourth expedition in 1931 when he was returning from having delivered supplies to friends who would have been stranded at their research post without enough food for the winter. His friends buried him in the snow and returned later to post a grave marker. Perhaps he is vindicated now, resting beneath the ice in Greenland as the entire North American plate crawls, centimeter by centimeter, toward the Pacific Ocean.

THE BIG QUESTION

Do scientists ever change their minds about how something on Earth works? What helped scientists eventually accept Wegener's claim?

Seafloor Spreading

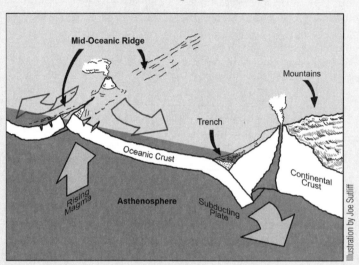

Wegener's theory of continental drift has developed into the modern theory of plate tectonics. The diagram above shows the movement of plates in the ocean and at the border of an ocean and a continent.

Questions

1. Arrows in diagrams can have different meanings. This diagram has two types of arrows. Draw a line to match the arrow type with what it is doing in the diagram.

Thin black arrows

Large white arrows

_____ pointing to something important that you should notice

_____ showing the direction that something is moving

_____ giving the name of an object in the picture

_____ showing that one thing turns into something else

2. Draw your own arrow on the diagram to label where the newest rock is forming.

3. What kind of rock would you expect to find in the mountains of a mid-ocean ridge (sedimentary, igneous, or metamorphic)?

4. Oceanic crust is denser than continental crust. What happens to oceanic crust when it meets continental crust?

5. In this diagram, what two features form when oceanic crust meets continental crust?

The Ocean on Top of a Mountain

Topics
- Geological dating
- Burgess Shale
- Mountain formation

Reading Strategy
- Finding the meaning of new words

Lesson Objectives: Connecting to National Standards

The following list shows the *Next Generation Science Standards* (*NGSS*) and *Common Core State Standards* (*CCSS*) supported by this activity.

NGSS: *Science and Engineering Practices*
- Developing and Using Models
- Constructing Explanations and Designing Solutions

NGSS: *Disciplinary Core Idea*
- **ESS1.C.** The History of Planet Earth

NGSS: *Crosscutting Concepts*
- Systems and System Models
- Patterns

CCSS: *Literacy in Science and Technical Subjects*
- **CCSS.ELA-Literacy.RST.6-8.4.** Determine the meaning of symbols, key terms, and other domain-specific words and phrases as they are used in a specific scientific or technical context relevant to grades 6–8 texts and topics.
- **CCSS.ELA-Literacy.WHST.6-8.2.** Write informative/explanatory texts, including the narration of historical events, scientific procedure/experiments, or technical processes.
- **CCSS.ELA-Literacy.WHST.6-8.2.c.** Use appropriate and varied transitions to create cohesion and clarify the relationships among ideas and concepts.

Background

Landforms develop and change over time, but they retain traces of their earlier structures. Geologists use those traces to piece together the story of the Earth's past. In this chapter, students will simulate the "life" of a mountain and then think about what traces of each stage of the mountain's life are present at the end of the simulation. Then they will read about the Burgess Shale and how scientists used the clues in the mountain to date the fossils and determine how marine fossils ended up near the top of a group of mountains.

SAFETY NOTE
Scissors and even shells could potentially cut someone. Use your good sense with these items. Have students wash hands after handling shells if the shells haven't been sanitized.

Materials (Per Group)

- 7–10 small shells (can be reused)
- 7–10 small shells that have been colored with a permanent marker (can be reused)
- 7–10 dried beans (can be reused)
- Colored construction paper (four colors, one or two sheets of each)
- Scissors
- Roll of tape or glue dots

Student Pages

- The Story of a Mountain (lab sheet)
- "Dating the Burgess Shale" (article)
- Reading the Rocks at the Grand Canyon (thinking visually)

Exploration/Pre-Reading

Before class, gather the materials for each group. You will need to color or dye some of the small shells to differentiate them. Begin by dividing students into groups of four. Groups can work through the simulation at their own pace.

Introduce the Reading. Tell students that they are going to read an article that will take them all the way back to the Cambrian era to explore a famous group of ocean fossils found on a mountain. The Royal Ontario Museum and the Yoho National Park (both in Canada) each have two- to three-minute videos that introduce the Burgess Shale. You could show one of them to set the stage for the article.

Reading Strategy: Finding the Meaning of New Words

Note that this strategy was first introduced in Chapter 5 (p. 45). If you have not used Chapter 5, introduce the strategy according to the instructions found on page 48 instead.

If your students have already been introduced to the strategy, begin by asking them if they can remember some of the ways that science books give definitions of new words. List their answers on the board and add any they may have forgotten (see Table 2.1, p. 16).

Next, explain that not all words are defined in the text. Sometimes a writer uses a word and doesn't give the definition. In those cases, you try to figure out a "good enough" definition. Post the following sentences:

An Anomalocaris was also on the hunt. It reached down with its strong, armored tentacles and grabbed the trilobite. The Anomalocaris tried to shove the struggling trilobite into its mouth.

Ask students what they think an *Anomalocaris* and a trilobite might be, and have them give the clues they are using. Here are some possible answers:

- *Anomalocaris* is some kind of animal because it is hunting.
- It might be like an octopus because it has tentacles. (*Note*: It's not actually related, but that is a reasonable guess based on the clues in the passage).
- *Anomalocaris* is a scientific name because it is in italics.
- Trilobites are something the *Anomalocaris* eats, because it is being shoved into the *Anomalocaris*'s mouth.
- Trilobites are also animals because this one is described as "struggling."

At this point, students probably don't have a clear picture of what an *Anomalocaris* is, but they know *enough* to keep reading. Ultimately, this article isn't about an *Anomalocaris*—it's about dating a fossil find. Therefore, the clues above tell them *enough* to understand the point of the article.

Note that there may be other words in the passage that students don't know, such as tentacles. See if they can figure out enough to keep reading (in this case, that tentacles are something an *Anomalocaris* has that lets it grab another animal). It can be very freeing to students to realize they don't really have to know a word completely to understand *enough*.

Journal Question

Do you ever have trouble remembering the meaning of a new word after you read it? What could you do while reading to help you keep up with the new words you are learning?

FIND OUT MORE
To learn more about the Burgess Shale, check out the Burgess Shale pages from the Royal Ontario Museum at *http://burgess-shale.rom.on.ca/en* or the book
- Morris, S. C. 1999. *The crucible of creation: The Burgess Shale and the rise of animals.* Oxford: Oxford University Press.

Application/Post-Reading

- Thinking Visually: Reading the Rocks at the Grand Canyon
- Writing Prompt: Write a paragraph that explains the difference between an absolute age and a relative age. Give an example of each from geology and then give an example of how you could describe how long you've been alive using a relative age and an absolute age.
  - Prewriting questions: What science words will you need? What writing words could you use? (compare and contrast words such as similarly, in contrast, however, etc.). You may need to help students brainstorm ways to give their relative age (older than one sibling, younger than another; older than the iPad, younger than cell phones, etc.).
  - Key Evaluation Point: Relative ages tell how old things are in comparison to each other, whereas absolute ages give dates for when a rock (or person!) was formed.

The Story of a Mountain

Decide which group member will hold each job. Write your job title at the top of your paper.

Jobs

Narrator: Reads the story that describes what happens to your model

Builder: Builds the land up

Eroder: Erodes and weathers your land

Fossilizer: Leaves traces of life

Work through the story below, with each group member doing their part. Answer the questions on your own paper.

Narrator: Once upon a time, there was an ancient ocean.

(**Builder**—*Lay one color of paper on the table to represent the ocean floor.*)

Narrator: Lots of different kinds of animals with shells lived in the ocean. When they died, their shells collected on the bottom.

(**Fossilizer**—*Sprinkle shells over the paper and tape them down.*)

Narrator: After several million years, the tectonic plate holding the ocean shifted. The water in this part of the ocean became shallower. Coarse rocks washed into the ocean and covered up the old ocean floor.

(**Builder**—*Place a different color of paper on top of the shells to represent the rock layer.*)

Narrator: At this time, there were some special animals living in the shallow edge of the ocean. They only existed for a few thousand years and then died out.

(**Fossilizer**—*Tape the colored shells onto the second paper.*)

Narrator: Early in Earth's history, there were many more volcanoes than there are now. One erupted near here, and lava covered what was left of the water in this area.

(**Builder**—*Add a third layer of paper to the pile to represent the lava layer.*)

Narrator: The lava filled up the ocean so it became land. Small land animals lived there. Some of their skeletons were preserved in the rock.

(**Fossilizer**—*Tape the beans onto the lava layer.*)

Narrator: An ice age hit. A big glacier slid across this area, stripping off some of the rock as it came through.

(**Eroder**—*Cut off half of the paper that was just put down, including any beans that are on it.*)

Narrator: Over the next several million years, bits of dirt and rock from a nearby mountain washed down in rainstorms over this area. Layers of sediment formed over the lava. Pressure converted the sediment into sedimentary rock.

(**Builder**—*Cover the top layer with the last piece of paper.*)

Everyone: *Pause to do the following on your own paper:*

1. Draw your landscape. Label the colors for each layer and where to find each type of fossil (shells, colored shells, beans).
2. Draw a squiggly line to show where some of the rock is "missing" due to the glacier. Here, the two layers that are next to each other don't follow each other exactly in time. Geologists call this an unconformity. Write "unconformity" next to your squiggly line.
3. Label the layer that holds the *oldest* rock and the *oldest* fossils.

Narrator: The tectonic plates are on the move. Two plates have collided, and this part of the Earth was squished.

(**Builder**—*Press the two sides of your model together until it folds up in the middle to form a mountain. Make a tight crease across the top of the "mountain" to make a fold that will stay.*)

Narrator: A mountain has formed! And almost immediately, a river forms, because water runs off the mountain when it rains. The movement of the water gradually wears down a valley in the middle of the mountain range.

(**Eroder**—*Use the scissors to cut across the middle of the crease at the top of the mountain. The cut should go through all the layers. Separate the two sides of the mountain so that you can see the inside.*)

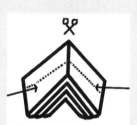

Everyone: *Pause to do the following on your own paper:*

4. Imagine you are on a raft on the river at the bottom of the valley. Draw what you would see looking up at one side of the mountain. Label the oldest and youngest layers, and the position of each group of fossils.

Narrator: The weather patterns in the area mean that erosion is worse in one area of the mountain. It happens to be the same place that the glacier went through millions of years before.

(**Eroder**—*Cut away part of the top layer of paper over the spot where you have an unconformity. You should now be able to see the layer with the colored shells.*)

Everyone: *Answer the following on your own paper:*

5. Imagine you are hiking this mountain with your friend and see one of the colored shells. She asks, "How did a fossil shell get on top of a mountain?" What do you tell her?

Dating the Burgess Shale

In the mud at the bottom of a tropical ocean, by the base of an underwater cliff, a squishy worm wiggled in its burrow. The wiggle was a mistake, though, because a trilobite was hunting nearby. The trilobite snatched the worm from its burrow and began to eat.

A shadow passed overhead. An *Anomalocaris* was also on the hunt. It reached down with its strong, armored tentacles and grabbed the trilobite. The *Anomalocaris* tried to shove the struggling trilobite into its mouth. But it was too slow.

Something disturbed the mud at the top of the cliff above them. Mud poured down in an underwater landslide, and everything, including the *Anomalocaris,* the trilobite, and the burrowing worm, was instantly smothered.

Artist's representation of *Anomalocaris* with a trilobite

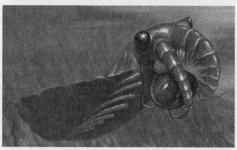

Fast Forward

Half a million years later, Charles Walcott, his wife Helena, and their 13-year-old son were walking through the Canadian Rocky Mountains. Walcott was a paleontologist, and they were on their way home from a fossil hunt. They stopped by a ridge between two peaks and broke open a rock. Walcott knew they had found something special. Usually, only the hard parts of animals become fossils. But here was the fossil of a squishy worm. Walcott looked around. The rocks had fossils of trilobites, *Anomalocaris,* and other animals with hard parts. But they were also full of the fossils of animals with soft bodies. He named the area the Burgess Shale. As soon as possible, he came back with a team to excavate the area.

Location of the Burgess Shale

To understand the fossils coming out of the Burgess Shale, scientists needed to know the story of the place where they were found. How old were the rocks and the fossils they contained?

Reading the Rock

The Burgess Shale fossils are preserved in sedimentary rock. It formed when layers of mud were compressed into rock. If an area of sedimentary rock hasn't been moved, then the oldest layers are at the bottom and the most recent layer is on top. By comparing layers, you can tell which rocks and fossils are older, but you can't tell exactly how old each layer is.

The oldest layers are at the bottom.

The rocks of the Burgess Shale have been moved by plate tectonics. A huge section of rock was shoved up and across North America as two plates collided, forming the Rocky Mountains. So the layers of the Burgess Shale are located partway up a mountain. However, within the slab of rock that was moved, the sedimentary layers go from oldest to youngest.

The fossils themselves can also provide clues to their age. Some fossils are found all over the world but only within a short range of time. For example, there are many fossils of the trilobite *Elrathina* found in North America. They all come from rocks formed during the Middle Cambrian period. Therefore, when new fossils of *Elrathina* show up, scientists can make a good guess that those rocks are also from the Middle Cambrian. Fossils that help determine the age of rocks are called index fossils. *Elrathina* is one of several index fossils found in the Burgess Shale.

Trilobite fossil from the Burgess Shale

5.0 mm

Making a Date

Sedimentary rock layers and index fossils help scientists determine if rocks are younger, older, or about the same age as other rocks. This is called a relative age. But to find out how many years old a fossil is, called the absolute age, scientists need an igneous rock. Igneous rocks contain small amounts of radioactive atoms. Radioactive atoms are unstable and break apart. Fortunately, they break apart at a steady, predictable pace. Scientists can test for how much radioactive material is left and figure out how long ago the rock formed. It's as if you bought a pack of gum and chewed one piece each day. You could figure out what day you bought the pack by looking at how many pieces were missing.

There are no layers of igneous rock in direct contact with the layers of the Burgess Shale. However, there are igneous layers some distance above and below the shale. The lower layer dates to 542 million years ago. The upper layer dates to 488 million years ago. So the Burgess Shale must be between 488 million and 542 million years old. By comparing the index fossils and sedimentary layers to other places where similar fossils and sedimentary layers have been found, geologists estimate that the Burgess Shale was formed between 505 million and 510 million years ago.

Life in the Cambrian Period

That date places the fossils firmly within the Cambrian period. Somehow, the mudslides from the underwater cliff had preserved soft and hard-bodied animals in more detail than anyone had seen before. These amazing fossils revolutionized our understanding of life during the Cambrian. The Burgess Shale is so rich in fossils that even after a hundred years of excavation, there are still more fossils, and more clues to life in the Cambrian, to be found.

THE BIG QUESTION

List three kinds of evidence from the article that scientists use to determine the age of layers of rock.

Reading the Rocks at the Grand Canyon

The diagram below shows the layers of rock that are exposed along the sides of the Grand Canyon in Arizona. Although all of the layers have names, only a few are labeled here. Use the diagram below to answer the questions.

Layers of the Grand Canyon

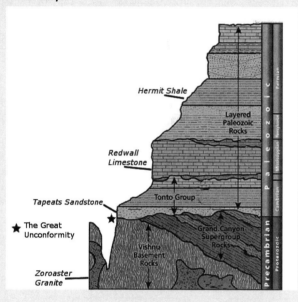

1. Four layers are listed in italics. Name those layers in order from youngest to oldest.

2. Considering the label "Great Unconformity," what can you tell about the rock layer just below the Tapeats Sandstone?

3. The layers below the Great Unconformity are tilted. What kind of forces could have caused them to tilt?

4. The Zoroaster Granite formed when magma pressed upward into the Vishnu Basement Rocks and then cooled into rock. Which must be older, the Zoroaster Granite or the Vishnu Basement Rocks?

Rock-Solid Evidence

Topics

- Igneous, metamorphic, and sedimentary rock
- The rock cycle

Reading Strategy

- Chunking

Lesson Objectives: Connecting to National Standards

The following list shows the *Next Generation Science Standards* (*NGSS*) and *Common Core State Standards* (*CCSS*) supported by this activity.

NGSS: *Science and Engineering Practices*
- Engaging in Argument From Evidence

NGSS: *Disciplinary Core Idea*
- **ESS2.A.** Earth's Materials and Systems

NGSS: *Crosscutting Concept*
- Energy and Matter: Flows, Cycles, and Conservation

CCSS: *Literacy in Science and Technical Subjects*

- **CCSS.ELA-Literacy.RST.6-8.1.** Cite specific textual evidence to support analysis of science and technical texts.
- **CCSS.ELA-Literacy.RST.6-8.9.** Compare and contrast the information gained from experiments, simulations, video, or multimedia sources with that gained from reading a text on the same topic.
- **CCSS.ELA-Literacy.WHST.6-8.1.** Write arguments focused on discipline-specific content.
- **CCSS.ELA-Literacy.WHST.6-8.1.a.** Introduce claim(s) about a topic or issue, acknowledge and distinguish the claim(s) from alternate or opposing claims, and organize the reasons and evidence logically.

Background

Many students have never looked carefully at rocks. If they have spent most of their time in cities, gravel may be their mental image of a rock. For this reason, looking closely at several different types of rocks is important for preparing them to learn about how rocks are classified. One goal of the article is to help them see how identifying rocks can be useful, both in forensics and in using matching rocks to understand the history of Earth. The definition of a mineral is not included in this lesson. You may wish to introduce minerals beforehand or use this lesson as a jumping-off point for exploring the minerals that make up rocks.

Materials (Per Group)

- 6 rock samples (with at least one each of igneous, sedimentary, and metamorphic), labeled S1–S6
- An extra piece of one of those samples (to serve as the unknown)
- Hand lenses and rulers (optional)

Student Pages

- "Rock-Solid Evidence" (article)
- Stolen Goods! (lab sheet)
- The Rock Cycle (graphic organizer)

SAFETY NOTE

Obviously, if you use rock samples that have sharp edges, do caution your students to be careful not to cut themselves. Students should wash their hands with soap and water after lab activities.

SAFETY NOTE

If students are testing rocks with a weak acid, they should wear nonlatex gloves, aprons, and eye protection (sanitized indirectly vented chemical-splash goggles meeting the ANSI Z87.1 standard) during the setup, investigation, and takedown. They should also wash hands with soap and water on completing the activity.

76

Exploration/Pre-Reading

Set the stage by inviting students to become forensic geologists, and explain that forensic geologists use their geologic knowledge to solve crimes. Geologists are often called in on "substitution" cases, in which a shipping container has been filled with rocks or sand instead of some expensive cargo. Finding the source of the rocks or sand is often a critical first step in identifying what happened to the cargo (or at least, whose insurance is responsible for paying for the fiasco!).

Give the students a moment to look over the samples. Then ask them to suggest characteristics that might be useful for identifying rocks. Ask why size isn't particularly important in making a match. Inform them that they need to be as thorough as they can in their descriptions to have sufficient evidence to match the unknown to one of the locations. Students are asked to consider the color, texture, and patterns in the rock. A fourth column is labeled "Other." If you would like them to test for reaction with a weak acid or measure density, you can have them put that information there. After students have examined all of the samples, provide them with the sample from the shipping container and have them make a claim as to where the substitution occurred.

Introduce the Reading. Tell students they are going to read an article that describes a real crime that forensic geology helped solve. Tell them to watch for how rocks from various places can be so different.

Reading Strategy

This strategy is first introduced in Chapter 6 (p. 55). If you have not used Chapter 6 with your students, explain that many sentences in science writing have a lot of ideas crammed together. It can be difficult to try to understand all the ideas at one time, but if students break the sentence down into chunks, they can think about each piece individually.

Place the following sentence on the board:

Other rocks are formed from bits of sand and clay and dirt, called sediment, that collect at the bottom of a river, ocean, or other low place.

If you have introduced chunking before, ask a volunteer to place slash marks (/) in the places where she or he might chunk the sentence, and work with the sentence in the chunks that the student selects. If you are introducing the strategy for the first time, place the slash marks yourself in the following locations:

> **EXTENSION**
> If your students are accustomed to writing about claims and evidence, help them include counterclaims in their response. Show them two small beakers filled with beans. One should include kidney beans and white beans; the other should hold only kidney beans. Hold up a kidney bean and ask if they could say for sure which beaker it came from (they can't). Point out that their written responses should include more than just what makes the unknown a good match with the location they have chosen; it should also include evidence that it doesn't match the rocks at the other locations.

Other rocks are formed from bits of sand, clay, and dirt, / called sediment, / that collect at the bottom of a river, ocean / or other low place.

Point out that the first segment of the sentence may be a little surprising if they haven't ever thought about how rocks are made. Ask if anyone saw anything that looked like it might have been bits of sand, clay, or dirt in the rocks they examined.

Then show the students that the phrase *called sediment* is an extra piece of information stuck into the middle of the sentence. Science writing frequently uses this kind of inserted phrase. Read the sentence, leaving out that phrase (i.e., *Other rocks are formed from bits of sand, clay, and dirt that collect at the bottom of a river, ocean, or other low place.)* to show that the sentence works without it. Then ask what information the inserted phrase adds. Sentences with inserted phrases are said to be an "interruption construction" and they can be confusing.

For the next phrase, ask how sand, clay, and dirt might collect at the bottom of a river or ocean. If no one knows, point out that water washes the material down. Finally, ask what kinds of low places might collect sediment (valleys, ditches, etc.).

Remind students that chunking a sentence is like eating a pie. People don't put a whole pie in their mouths at one time: everyone eats it bite by bite, and some people need to take smaller bites than others. Similarly, some people will need to break a sentence into more chunks than others, and that is okay. For this article, students can separate the chunks using slashes, as you did on the board. When they are reading something they aren't allowed to write on, they can chunk it in their head or cover up the parts of the sentence that they aren't thinking about.

Journal Question

Do you ever "chunk" sentences when you are reading and think about them one idea at a time? Think about the kinds of things you have to read for school. Which ones would be easier to read using a chunking strategy?

Application/Post-Reading

- Graphic Organizer: The Rock Cycle
- Writing Prompt: (Have student revisit two or three of the rock samples from the lab. Choose those with clear characteristics such

as striations for metamorphic rock or obvious cobbles in sedimentary rock.) On the basis of what you read in the article "Rock-Solid Evidence," decide if each sample is igneous, sedimentary, or metamorphic rock. Describe the characteristics that make it fit that grouping.

○ Prewriting Questions: What characteristics might tell you a rock is igneous? sedimentary? metamorphic? What science words will you want to include?

○ Key Evaluation Point: At this point, students will only be able to identify rocks by a few characteristics. Depending on the samples, students might identify a rock as igneous if it is glassy or if it has grains of several colors of minerals. A sedimentary rock may have visible cobbles or have obvious layers. The only metamorphic rocks they could identify from the information in the article would be those with striations.

> **FIND OUT MORE**
> To learn more about forensic geology, see
> • Murray, R. C. 2011. *Evidence from the Earth: Forensic geology and criminal investigation.* 2nd ed. Missoula, MT: Mountain Press.

Rock-Solid Evidence

The town of Sioux Falls, South Dakota, was in shock over the murder of a nine-year-old girl. Her body had been found in a remote wooded area outside of town. The police had a suspect, and they had rock and soil samples from his car and from the place the body was found.

That's when they called in geologist Jack Wehrenberg, who specialized in identifying rocks and dirt from crime scenes. He looked to see if the bits of rock from the suspect's car would connect the suspect with the killing.

A Rock Is a Rock?

At first glance, rocks just look like rocks. But on closer inspection, differences appear. Some rocks have spots or stripes of different colors. Some are hard, whereas others may chip easily or crumble. Some resemble clumps of dirt, whereas others can be smooth and glassy.

On the Trail

Dr. Wehrenberg looked over the sample from the wooded area. There were some unusual bluish mineral crystals. He turned to the samples from the fender of the suspect's car and found more bluish crystals. Could they have come from the same rock?

Igneous Rock: Born in Fire

Geologists like Wehrenberg divide rocks into three main categories, based on how the rocks formed. Rocks that form from a volcano's hot lava or from magma deep within the Earth are called igneous rocks. If the magma or lava cools quickly, smooth glassy rocks can be born. If the magma cools slowly, the minerals have time to grow into crystals. The more slowly the magma cools, the larger the crystals grow. Figure S8.1 shows a chunk of diorite with both dark- and light-colored crystals.

Figure S8.1. Diorite With Both Dark- and Light-Colored Crystals

├─1 INCH─┤

Sedimentary Rock: Bits of Bits

Other rocks are formed from bits of sand, clay, and dirt, called *sediment*, that collect at the bottom of a river, ocean, or other low place. Over millions of years, the pile of sediment gets deeper and weighs heavily on the bottom layers. Water dissolves some of the minerals, and those minerals become like glue that sticks the layers of sediment together. Geologists call these glued-together bits of old rocks sedimentary rocks. In Figure S8.2, you can almost see the individual grains of sand that make up this sedimentary rock, called *sandstone*.

Figure S8.2. Sandstone

Metamorphic Rock: A Complete Makeover

The third category of rock includes rocks that start out as either igneous or sedimentary rocks. They are buried deep in the Earth and subjected to intense heat and pressure. The heat and pressure isn't quite enough to melt the rock back into magma, but it can cause the rock to soften and bend. Some of the minerals may loosen their grip on each other and rearrange inside the rock. Oftentimes, similar minerals group together, giving the rock a striped appearance, as shown in Figure S8.3. Such rocks that have been changed by heat and pressure, are called *metamorphic rock*, from the Greek word *metamorphosis*, which means "to change."

Figure S8.3. Metamorphic Rock

Infinite Variety

Within each category of igneous, sedimentary, and metamorphic rock, there is an almost infinite number of combinations. Magma varies in what liquid elements it contains, and each combination creates distinctive rocks. Sedimentary rock can be formed from any combination of other rocks, broken and pieced back together. Metamorphic rock can take any of the huge range of igneous, sedimentary, and even other metamorphic rocks and turn them into something new. As a result, rocks that formed in any place and time together are often different from rocks anywhere else on Earth. Even rocks with a shared name, such as granite, can vary slightly in their exact mineral composition.

This lets scientists match rocks from mountains in one part of the world with rocks from mountains across the ocean and know that at some point the mountains must have been joined. Likewise, it lets geologists such as Dr. Wehrenberg match rocks from a crime scene to a suspect and know that they were also together at some time.

A Solid Case

The blue crystals from the two samples were large. Dr. Wehrenberg recognized them as gahnite crystals. Gahnite forms from zinc, aluminum, and oxygen in metamorphic rocks. But gahnite is rare and had never been identified in South Dakota. Dr. Wehrenberg wasn't sure how rock containing gahnite ended up in South Dakota. But however it got there, it was with both the victim and the suspect. On the basis of Dr. Wehrenberg's testimony and other evidence, the suspect was found guilty and eventually confessed.

THE BIG QUESTION

List and describe the three major categories of rocks.

Stolen Goods!

A truck at a perfume factory was loaded with 2 million dollars' worth of exotic perfumes. The truck took the load to a freight train a few miles away. The train made two stops before reaching the ocean, where the boxes were loaded on a boat. When that boat reached its destination, the boxes were opened. Instead of perfume, the boxes were filled with rocks. The buyer was furious!

You've been called in to help solve the crime. Police need to know where the perfume was replaced with rocks, so they will know where to focus their search. You begin by collecting rock samples from along the route shown in Figure S8.4.

Figure S8.4. Route Taken by the Perfume Truck

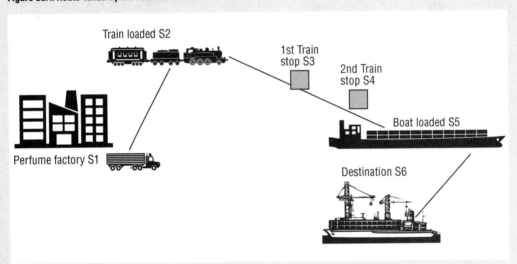

Use your data to make and support a claim, answering the following questions:

Claim: Where should police focus their search?

Evidence: Describe the rock data that support your claim.

Reasoning: Explain how that evidence shows where the police should start looking.

Sample	Color: What color(s) do you see in the rock? Are there spots or stripes?	Texture: Is the rock smooth, rough, glassy, crumbly, etc.?	Clumps or Layers: Describe the different parts that make up the rock.	Other
S1				
S2				
S3				
S4				
S5				
S6				
Sample from perfume box				

The Rock Cycle

Figure S8.5 shows how igneous, sedimentary, and metamorphic rocks are related. Study the diagram and decide which rock type belongs in each blank. Then answer the questions below.

Figure S8.5. Relationships Among Igneous, Sedimentary, and Metamorphic Rocks

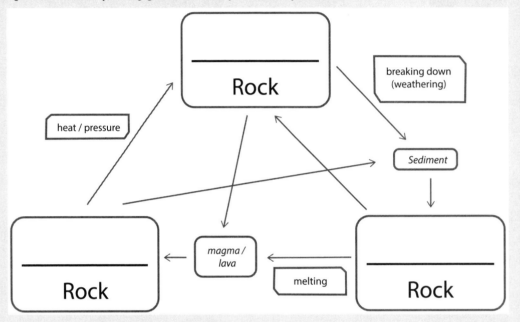

Questions

1. Arrows in diagrams can have different meanings. What are the arrows on this diagram doing?

    A. Pointing to something important that you should notice
    B. Giving the name of an object in the picture
    C. Showing that one thing turns into something else

2. Compare the information in this diagram with the information in the article "Rock-Solid Evidence." What information can you get from this diagram that you can't get from the article?

Look Out Below!

Topics

- The water cycle
- Groundwater
- Karst topography

Reading Strategy

- Talk your way through it

Lesson Objectives: Connecting to National Standards

The following list shows the *Next Generation Science Standards* (*NGSS*) and *Common Core State Standards* (*CCSS*) supported by this activity.

NGSS: *Science and Engineering Practice*
- Developing and Using Models

NGSS: *Disciplinary Core Ideas*
- **ESS2.A.** Earth's Materials and Systems
- **ESS2.C.** The Roles of Water in Earth's Surface Processes

NGSS: *Crosscutting Concepts*
- Cause and Effect
- Energy and Matter: Flows, Cycles, and Conservation
- Systems and System Models

CCSS: *Literacy in Science and Technical Subjects*
- **CCSS.ELA-Literacy.RST.6-8.2.** Determine the central ideas or conclusions of a text; provide an accurate summary of the text distinct from prior knowledge or opinions.
- **CCSS.ELA-Literacy.RST.6-8.9.** Compare and contrast the information gained from experiments, simulations, video, or multimedia sources with that gained from reading a text on the same topic.
- **CCSS.ELA-Literacy.WHST.6-8.2.** Write informative/explanatory texts, including the narration of historical events, scientific procedures/experiments, or technical processes.

Background

This lesson assumes the students have had some encounter with the water cycle previously, usually this happens in elementary school. However, many students do not understand the important role of groundwater in the cycle (Cheek 2010). Karst terrain provides a hook for studying groundwater, for thinking more deeply about the effects of the water cycle, and for tying this topic to other topics in geology.

Materials (Per Group)

- 1 clear cup
- 1 drinking straw (You may need to trim the straws so that the tube from the pump top reaches the water level in your cups.)
- Tape
- 1 cup aquarium gravel
- 1 pump top from a soap or hand sanitizer bottle
- 1 disc of green felt, same diameter as cup
- 1 pitcher of water
- Limestone rock sample (those with fossil shells are particularly good)
- White vinegar or 1 M HCl (hydrochloric acid)

- 1 dropper
- Hand lenses

Student Pages

- Water Underfoot (activity sheet)
- "Look Out Below! Sinkholes and the Water Cycle" (article)
- Practicing Talk Your Way Through It (modified reading group procedure)
- The Water Cycle (thinking visually)

Exploration/Pre-Reading

Students will begin with a two-part activity. In Part 1, they make a model of groundwater and think about the role of groundwater in the water cycle. In Part 2, they examine limestone samples and look for the characteristic carbon dioxide bubbles that are released when limestone reacts with an acid. You may use vinegar instead of hydrochloric acid, but the reaction will be milder. They may need to look closely to see the bubbles. You can reduce the amount of materials needed if some groups begin with Part 1 and others with Part 2. At the end, students are asked to hypothesize how the two parts might be related.

There are several dramatic video clips of sinkholes online. You may wish to show one of these clips before introducing the reading.

Introduce the Reading. Tell students that they are going to read an article that talks about a surprising connection between limestone and groundwater. As they read, they can see if the guess they made on their lab sheet is correct.

Reading Strategy: Talk Your Way Through It

Project the following paragraph from the reading and have a student read it aloud:

If you urinated this morning, drank a glass of water, or washed your hands, you participated in the water cycle. The water cycle is the phrase used to describe the movement of water around the Earth. You're probably familiar with some of these, such as rain, snow, or other precipitation falling from the atmosphere. When that precipitation reaches the ground, some of it runs into rivers, lakes, and oceans. Some of it ends up frozen in ice caps or glaciers. And some of it soaks into the soil. Water that soaks into the ground is called groundwater. And groundwater is more powerful than you might imagine.

Say, "There was a lot of information in that paragraph. Sometimes, it is hard to remember everything when you read a lot of new information at one time. One way to remember more and make sure you understand what you are reading is to pause and talk yourself through what you just read."

Then model for students how to use the strategy of talk yourself through it (also called pause, retell, and compare). For example, you might say, "Okay, so from the first reading of this, I remember that rain is part of the water cycle, that it ends up in oceans and lakes, and that groundwater is for some reason powerful."

Ask the class to look back at the paragraph and see what else you should remember from it. Tell students that by pausing to talk yourself through dense passages, you can make sure you understand the current paragraph before moving on to the next one.

Other paragraphs in this article are also dense and may contain information students have not learned before. Instruct them to try pausing after each section in the text to talk through (out loud, silently, or on paper) what they just read before moving on. If students are using reading groups, this prompt will be a part of their procedure, "This part was telling us that …"

Journal Question

Some people think that good readers don't have to use a strategy such as talk yourself through it. In fact, most people who remember what they read use a strategy like this. Did you find it easier to remember what you read after trying this strategy? Why or why not?

Application/Post-Reading

- Thinking Visually: The Water Cycle
- Writing Prompt: Late one night in 2011, a UPS store in Georgetown, South Carolina, collapsed into a sinkhole. The store was located over a limestone aquifer that had been stable until the collapse. The owners are suing the South Carolina Department of Transportation because the department had been draining groundwater

nearby to allow for the installation of some underground structures. Pretend you are filing a legal brief (report) to support the UPS store's case. Explain how the Department of Transportation's groundwater work might have led to a sinkhole under the UPS store.

- o Prewriting Questions: Will you want to use formal language or everyday language in this writing? What science vocabulary will you want to include? You will be writing about a process. What writing words might you use as you describe it? (*first, next, finally, therefore,* and *in conclusion* are all possibilities)
- o Key Evaluation Point: The answer to this question should include that limestone can dissolve in acidic water, but that, as long as the groundwater is in place, the remaining mineral structure may be fine. When groundwater is removed, a collapse is likely.

Reference

Cheek, K. A. 2010. Commentary: A summary and analysis of twenty-seven years of geoscience conceptions research. *Journal of Geoscience Education* 58 (3): 122–134.

Look Out Below!

Water Underfoot

Part 1: Groundwater Model

1. Tape the straw against the inside of your plastic cup. Leave about 1 cm between the end of the straw and the bottom of the cup. This will be your well.
2. Pour gravel into the cup until it is about three-quarters full. This represents rock below the surface of the ground.
3. Add water until the water line is just below the level of the gravel. This is groundwater, or water that is below the ground.
4. Place a piece of tape on the outside of the cup to show the water level. This line marks the water table.
5. Place the felt on top of your rocks. This is the surface material: all of the dirt, grass, and plants that are on top of the underlying rock. Your model should now look like Figure S9.1.

Figure S9.1. Groundwater Model Setup

Illustration by Joe Sutliff

You now have a model to show how water can exist below the ground. Use it to answer the following questions:

1. You filled the cup with rocks. How was there room for water in the same space?

2. Slide the pump into your straw, and start pumping into your extra cup. This represents drawing water up through a well for people to use. What happens to the water table as you pump?

3. If this model were real, what would need to happen to bring the water table back up?

90

4. Pour water into your cup to simulate rain. How does the water get into the ground?

5. Draw your groundwater model. Label the groundwater, water table, well, and surface materials.

Part 2: Limestone Rock

Safety Note: Be sure to wear the safety equipment your teacher has provided!

1. Look closely at the limestone rock sample. Use words to describe what you see.
2. Draw the limestone sample.
3. Put on your goggles. Place three drops of weak acid (either vinegar or weak HCl) on the limestone rock. Do you see bubbles? (You may need to look through a hand lens.)

Take a guess: How could groundwater and limestone rock be related to each other?

Look Out Below! Sinkholes and the Water Cycle

If you urinated this morning, drank a glass of water, or washed your hands, you participated in the water cycle. The water cycle is the phrase used to describe the movement of water around the Earth. You're probably familiar with some of the movements, such as rain, snow, or other precipitation falling from the atmosphere. When that precipitation reaches the ground, some of it runs into rivers, lakes, and oceans. Some of it ends up frozen in ice caps or glaciers. And some of it soaks into the soil. Water that soaks into the ground is called groundwater. And groundwater is more powerful than you might imagine.

One of the largest sinkholes in the United States is shown.

Florida Drop Zone

In 2013, guests at the Summer Bay Resort near Disney World were awakened by a security guard ordering them to evacuate the building. They all got out, and not a moment too soon. Within 15 minutes, most of their lodge had collapsed into a hole. In Mulberry, Florida, in 1994, a sinkhole destroyed a toxic waste disposal facility. The sinkhole released 20.8 million pounds of phosphoric acid into the surrounding area and created a pit about 60 m (200 ft.) deep. One of the most famous Florida sinkholes occurred in Winter Park in 1981, taking with it a house, a community swimming pool, a car repair shop, and five Porches that the shop was supposed to be fixing.

The culprit behind these collapses? Interactions between the water cycle, rocks, and human activity. Each of these tragedies took place on a type of geologic formation called karst terrain. In areas of karst terrain, the underlying rock can dissolve in water. The result, shown in Figure S9.2, is a unique and often beautiful landscape. Karst terrain can hold amazing networks of caves, such as Mammoth Cave in Kentucky. It may hold miles of lakes where water has settled into areas that have sunk. It may have disappearing rivers that drop suddenly below the ground, leading to gorgeous underground lakes. The very "holefull" nature of karst terrain also makes it a good storage tank for water. Places where water gathers underground are called aquifers. Aquifers that are in karst terrain provide freshwater for people all over the world. However, karst can also be dangerous.

Figure S9.2. Karst Terrain

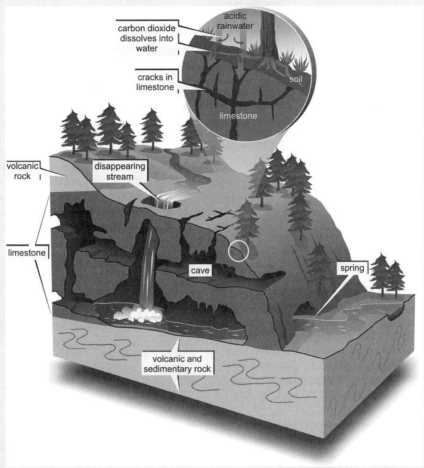

The Hole Story

Limestone is the most common rock under-lying karst terrain. Limestone is a sedimen-tary rock composed of calcium carbonate and other minerals. It often forms from the shells and skeletons of ocean animals. Over hundreds of thousands of years, these animal parts collect on the bottom of the ocean. Eventually, pressure joins all of the pieces into limestone.

Calcium carbonate is a base. When it is exposed to even a weak acid, a chemical reaction begins. That reaction slowly dissolves the calcium carbon-ate and releases tiny bubbles of carbon dioxide gas. Rainwater is often slightly acidic, and it becomes groundwater that is slightly acidic. When you have acidic groundwater in an area of limestone, you have a sinkhole just waiting to happen.

Completing the Cycle

Human decisions play a role in sinkholes too. Remember that water you used this morning? It had to come from somewhere.

In many places, people tap into groundwater supplies to get water for drinking, flushing, or even irrigating farms. After the water is used, it finds its way into lakes and oceans.

The energy from the Sun drives the water cycle by heating and evaporating surface water. The water vapor rises and collects in the atmosphere. It's cold up there, so the water condenses into tiny droplets that form clouds. Those droplets combine until they fall as precipitation, pulled down by the basic force of gravity.

As long as people use the groundwater at the same rate that it refills from precipitation, the ground over karst terrain has a good chance of remaining stable. Even as the calcium carbonate slowly dissolves, the groundwater itself helps support the framework of minerals left behind. But if people draw off too much groundwater, the water level drops. There's no longer anything to support the weakened rock, and the risk of a sinkhole rises. The ground may sink down slowly over years, or it may cave in suddenly.

Sinkholes can also be formed when extra weight is placed on top of unstable rock. A sinkhole formed in Guatemala City when tremendous rains from a tropical storm pooled around a three-story building. The ground could not support the weight of the water and the building. A 30 m (100 ft.) sinkhole swallowed the entire structure.

Why Florida?

Karst terrain is found all over the world. There is some in almost every state in the United States. Florida, however, is especially prone to dramatic sinkholes because almost all of Florida sits on limestone. The population of Florida is increasing. More people means more water must be drawn off to meet everyone's needs. More people means that more structures are built that could overload the fragile terrain. And more people means there is a good chance that someone will be around to witness the next time the ground collapses.

Many lakes in Florida formed from sinkholes, as shown in this satellite image.

THE BIG QUESTION

In your lab, you were asked to guess how limestone and groundwater could be related. Based on the article, describe at least one relationship between limestone and groundwater.

Practicing the "Talk Your Way Through It" Strategy

Group Reading Procedure

1. Everyone reads the first section quietly and marks !, ✓, ×, and ? while reading.
2. The *leader* describes what the section was about. If you're stuck, try starting with the phrase, "What I understand so far is …"
3. The leader asks each member of the group to share anything that was confusing (marked ? or ×).
4. The group should try to figure out what the confusing word, sentence, or idea means. If the group cannot make sense of the confusing word, sentence, or idea, the *emergency manager* should raise the flag to get help from the teacher.
5. Repeat steps 1–4 for Section 2 (separated by a line), but this time the *interpreter* describes what the section was about.
6. Repeat steps 1–4 for Section 3 (separated by a line), but this time the *emergency manager* describes what the section was about.
7. The group should work together to answer the Big Question. The *interpreter* will write the group's answer to turn in to the teacher.

The Water Cycle

On the drawing below, add arrows and labels to show how water moves around the Earth. You may wish to look back at the article, "Look Out Below!" to ensure that you have included all of the ways that water moves and is stored. Add information to the drawing that shows how gravity and the Sun's energy drive the cycle.

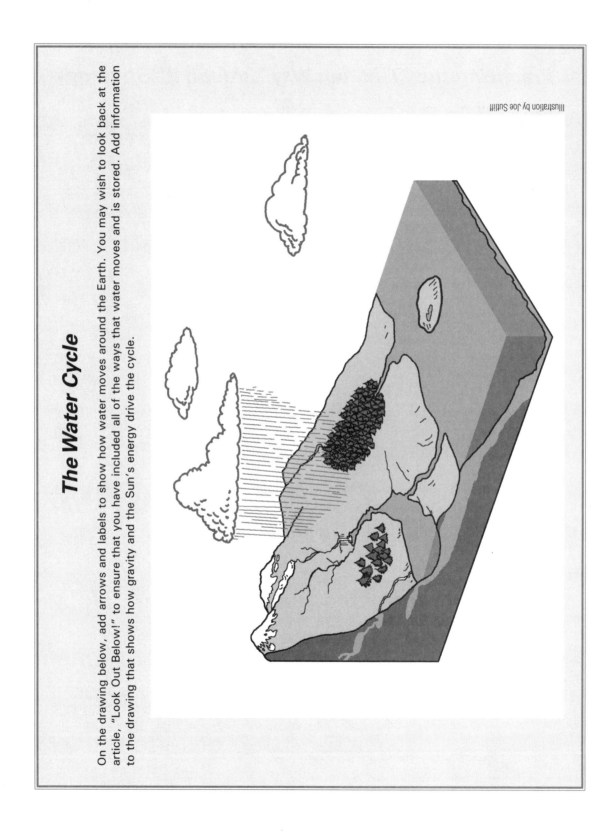

Illustration by Joe Sutliff

Oceans on the Move

Topics

- Deepwater ocean circulation
- Convection
- Density
- Using models in science

Reading Strategy

- Previewing diagrams and illustrations

Lesson Objectives: Connecting to National Standards

The following list shows the *Next Generation Science Standards (NGSS)* and *Common Core State Standards (CCSS)* supported by this activity.

NGSS: *Science and Engineering Practice*
- Developing and Using Models

NGSS: *Disciplinary Core Idea*
- **ESS2.C.** The Roles of Water in Earth's Surface Processes

NGSS: *Crosscutting Concept*
- Systems and System Models

CCSS: *Literacy in Science and Technical Subjects*
- **CCSS.ELA-Literacy.RST.6-8.7.** Integrate quantitative or technical information expressed in words in a text with a version of that information expressed visually (e.g., in a flowchart, diagram, model, graph, or table).
- **CCSS.ELA-Literacy.RST.6-8.9.** Compare and contrast the information gained from experiments, simulations, video, or multimedia sources with that gained from reading a text on the same topic.
- **CCSS.ELA-Literacy.WHST.6-8.2.** Write informative/explanatory texts, including the narration of historical events, scientific procedures/experiments, or technical processes.

Background

The current model of deep ocean circulation is in flux, as data over the past few decades have shown the "global conveyor belt" model to be inaccurate. This provides a great opportunity to explore the idea of models and how models change with new information. To complicate the discussion, however, many middle school students still struggle with the concept of density, without which it is difficult to understand overturning circulation. Therefore, this chapter begins with physical models of temperature and saline density differences and moves to looking at the work of a researcher building computer models of overturning circulation.

This is a timely topic. Ocean currents move heat around the Earth and alter climates. They also serve as a carbon dioxide sink, or a trap for carbon dioxide. Research in this area has been spurred by concerns about what might happen to currents if surface temperatures rise, and what might happen to carbon dioxide storage if currents change. Scientists are trying to find out if warming temperatures could set off a feedback loop in which the ocean releases carbon dioxide, or doesn't absorb it as well, leading to more warming.

Materials

For convection current:
- 2 ft. of ¼-in. clear vinyl tubing (available in plumbing section of a hardware store)
- 1 T-connector for ¼-in. vinyl tubing
- 1 thin pipette
- Food coloring
- 1 microwaveable heat pack

For saline density:
- 1 clear plastic shoe box
- Aluminum foil
- Clear tape
- Sharp pencil
- Salt
- Water
- Red and blue food coloring

Student Pages

- Model Mania (lab sheet)
- "Oceans on the Move" (article)
- Convection Currents (thinking visually)

Exploration/Pre-Reading

There are many ways to model convection and density differences. Two models are described next, but feel free to substitute your favorite models.

Convection Current. Science supply companies sell glass convection tubes that can be heated at the corner with a Bunsen burner. These provide an excellent demonstration but can be expensive. Similar, but a little slower, results can be accomplished with a homemade substitute made from the materials in the supply list above.

Before class, prepare the equipment by connecting the tubing with the T-connector, leaving the unused connection pointing up. Fill the tube with

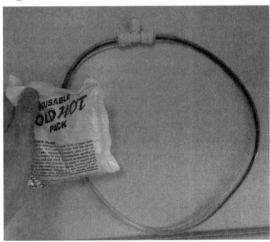

Figure 10.1. Convection Model Setup

water from a faucet, removing as much air as possible. Tape the tubing to a white surface, as shown in Figure 10.1, p. 99. Heat the microwaveable heat pack according to the instructions on the package. You may wish to hold it with a heat resistant glove.

When you are ready to do the demonstration, place one drop of food coloring into the top of the T-connector. You will need to use a thin pipette to insert the drop all the way into the bottom of the connector. You should see a small amount of color leak into the tubing. Have students observe that there is not much movement at this point. Ask them to predict what will happen if you heat one side of the tubing.

Next, hold the heat pack against one side of the tubing. In 30 seconds to a minute, the food coloring should begin to move away from the side with the heat source. Allow students to see it move about a quarter of the way around the circle. Then move the heat source to just under the food coloring. In a moment, the color will reverse its movement. This time, allow students to see the coloring to complete the convection cycle.

Figure 10.2. Salinity Separating the Red and Blue Water

Saline Density. Prepare the shoe box ahead of class by taping aluminum foil across the inside and onto the bottom so that it separates the box into two sections. When you are ready for the demonstration, fill two containers with water. Mix as much salt as you can into one of the containers and dye the water red. Dye the unsalted water blue. Pour one solution into each side of your container at the same time; this helps keep the pressure from tearing the aluminum foil. Ask students to predict what will happen if you poke two holes in the foil. Use the sharp pencil to poke one hole just below the surface of the water and the other near the bottom of the container. Observe the flow for two to three minutes. You can see the end result in Figure 10.2.

Introduce the Reading. Tell students that the activities you have done are models for some of the variables that affect the movement of water in the ocean. Tell them they are going to read about a scientist who is studying those movements.

Reading Strategy

Tell students that in some books they read, the pictures may be extras. In science writing, however, the pictures and diagrams often carry a lot of important information. Looking at the pictures and making predictions about what they mean before reading can help make the text easier to understand.

Hand out the article, "Oceans on the Move," but tell students not to read it yet. Look through the figures as a class and provide students with the following prompts to help them study the diagrams and predict their meanings. Place students in pairs or in their reading groups to discuss the prompts. For the prediction questions, accept any answer as a valid possibility and tell students that they will have to read the text to find out if they are right.

Start with Figure S10.1. Ask the following questions:

- What do you think the circles and lines represent in this diagram? (water molecules) What are the arrows showing? (movement)
- What do this picture and caption indicate about hot and cold water?
- How do you think this diagram could be related to what happened in our first demonstration? (prediction)

Move to Figure S10.2. Ask the following questions:

- What is this diagram showing?
- What do the balls and lines represent?
- Can you think of any way this could be related to our second demonstration? (prediction)

Look at Figure S10.3. Ask the following questions:

- What are the arrows showing in this picture (movement of wind and water)?
- Based on this picture, what do you think the word "upwelling" means?
- How do you think this is related to ocean currents? (prediction)

Finally, turn to Figure S10.4. Ask the following questions:

- What do you see in this picture?
- How could this relate to ocean currents? (prediction)

Journal Question

Do you typically study the diagrams and pictures in your science book? Was it helpful to look at the diagrams and pictures before reading today? Why or why not?

FIND OUT MORE

To learn more about recent developments in ocean circulation, see

• Lozier, M. S. 2010. Deconstructing the conveyor belt. *Science* 328 (5985): 1507–511.

Application/Post-Reading

- Thinking Visually: Convection Currents
- Writing Prompt: Revisit the second model from your lab activity. Using information from the article, explain why the water moved as it did and how this model relates to ocean circulation. Feel free to include diagrams in your explanation.
  - Prewriting Questions: Make a quick list of the key ideas you want to include. Number them in the order you will use them. What science words will you want to include? You will be writing about a process that involves cause and effect. What kinds of writing words could help you structure your explanation? (*first, next, then, because, therefore*)
  - Key Evaluation Point: In the model, the salt water moved across the bottom because the salt made it denser. The freshwater spread across the top because it was less dense. As water freezes in the ocean, the salt is left behind. The remaining water is left with the salt, which sinks because of its density.

Model Mania

Your teacher will conduct two demonstrations. For each demonstration, draw and label a diagram of the setup and answer the questions that follow. Show heat and/or salt and the direction the water moved.

Model 1 (Draw It):
• What do you predict will happen when your teacher heats the tubing?

• Describe what actually happened.

• Why do you think the colored water moved as it did?

Model 2 (Draw It):
• What do you predict will happen when your teacher punches holes in the aluminum foil?

• Describe what happened when the holes were punched:

• Why do you think the water moved as it did?

Both of these demonstrations are examples of models. Models help scientists understand how the world works. They can also help scientists predict what will happen in new situations. These two models explore variables that affect how ocean water moves.

• What variable does the first model test?

• What variable does the second model test?

Oceans on the Move

When Dr. Susan Lozier looks across the ocean from the deck of a research boat, she wonders where all that water has come from, and where it is going. Scientists have known since the 1800s that ocean water doesn't stay in the same area forever. The water that is now crashing against the beaches of the United States will gradually find its way around the world. It will also travel from the surface of the ocean to the depths and back again. The currents that move the water up and down and around the world take part in what is called overturning circulation.

Through overturning circulation, ocean water carries nutrients, oxygen, carbon dioxide, and heat around the world. Therefore, it is critical to life on Earth.

For many years, scientists had a model of how they thought overturning circulation worked. Lozier and other researchers have been testing the model with high-tech floats. These floats travel with the currents underwater. After two years, they rise to the surface and beam their data to satellites.

However, the floats keep popping up in places where they aren't expected. This raises a big question. If the old model is wrong, how *does* water move around the world? Scientists know some of the main causes.

What Factors Cause Ocean Water to Travel?

Heat. Water looks like a continuous substance. But like all matter, water is made up of atoms and molecules. When water is a liquid, the molecules move around in a jumble. In cold water, the molecules move slowly and stay close together. But when heat energy is added, the water molecules move faster and spread out. There are actually more molecules in a liter of cold water than there are in a liter of hot water (Figure S10.1). Indeed, a liter of water just above freezing weighs about 40 g more than a liter of water just below the boiling point. The scientific way to describe this difference is to say that cold water is denser than hot water.

Figure S10.1. Molecules in Warm and Cold Water

Molecules in warm water (A) move quickly and spread out. Cold water molecules (B) move more slowly and pack more densely.

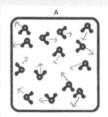

When two fluids mix, the denser substance sinks to the bottom. Therefore, when hot and cold waters meet, the colder, denser water sinks to the bottom. Ocean currents driven by differences in temperature are called convection currents.

Salt. As ocean water moves into the polar regions, some of the water freezes. The water molecules in ice line up to make perfectly shaped crystals, as shown in Figure S10.2. There is no room for salt in the ice, so ocean water that freezes leaves its salt behind. That means that the ocean water that does *not* freeze is extra salty. Those salt molecules pack in with the water molecules and make the water denser.

Figure S10.2. Water Molecules in Ice Fitting Neatly Together

This means that polar waters face a density double whammy: they are both cold and salty. They sink below warmer, less salty water.

Eddies. Currents push past islands and the edges of continents. They also encounter pockets of water with different densities. These and other factors can cause a swirl of water called an *eddy*. Eddies blend the deeper cold water with warmer surface water and affect the path the water takes.

Wind. Strong winds can push the surface water so that it moves in one direction. As shown in Figure S10.3, deep water then moves up to take the place of the surface water. This is one way that nutrients that have sunk to the bottom of the ocean come back to the surface.

The Model Makers

It's one thing to know the variables that affect overturning circulation. It's another

Figure S10.3. Molecules in Warm and Cold Water

Wind moves the surface waters. Because the Earth is turning, the water eventually moves at a right angle to the wind.

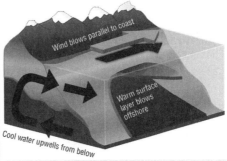

to figure out what the water is doing when all of those variables are working together. Researchers use computer models to show what role each variable might play (Figure S10.4). They also use their models to predict what might happen in the future if a variable changes. Researchers are particularly concerned about predicting how ocean circulation could change as the Earth warms.

There are about two dozen models for ocean circulation right now. And they are all different.

Figure S10.4. Dr. Susan Lozier in the Lab on a Research Ship

"What we need," says Lozier, "is more real-world data." She's already working with scientists across the Northern Hemisphere to track more floats through the Atlantic Ocean. Data from those floats will help Lozier and her colleagues settle on a new model for where the water has come from and where it is going.

THE BIG QUESTION

Scientists often create models to show how they think something works. Do scientists ever change these models? Provide an example from the text to support your answer.

Convection Currents

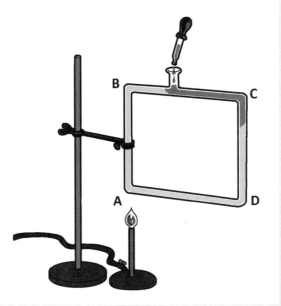

Illustration by Joe Sutliff

1. Which letter in the image above indicates the corner where the water is hottest?
2. When water is heated, what happens to its density?
3. Draw an arrow to show the direction the water will move between A and B.
4. As the water moves away from the flame, what happens to its temperature?
5. What happens to its density as it moves away from the flame?
6. Draw an arrow to show the direction the water will move between C and D.

Convection currents also take place in gases. Draw arrows to show how you would expect the air to move in this room.

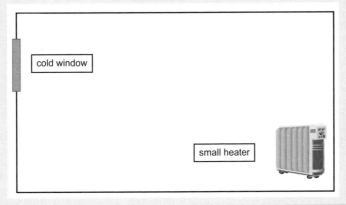

Trash Soup

Topics

- Global wind patterns and currents
- Coriolis effect
- Ocean garbage patches

Reading Strategy

- Identifying text signals for cause and effect

Lesson Objectives: Connecting to National Standards

The following list shows the *Next Generation Science Standards (NGSS)* and *Common Core State Standards (CCSS)* supported by this activity.

NGSS: *Science and Engineering Practice*
- Constructing Explanations

NGSS: *Disciplinary Core Ideas*
- **ESS2.C.** The Roles of Water in Earth's Surface Processes
- **ESS3.C.** Human Impacts on Earth Systems

NGSS: *Crosscutting Concept*
- Scale, Proportion, and Quantity

CCSS: *Literacy in Science and Technical Subjects*
- **CCSS.ELA-Literacy.RST.6-8.1.** Cite specific textual evidence to support analysis of science and technical texts.
- **CCSS.ELA-Literacy.RST.6-8.5.** Analyze the structure an author uses to organize a text, including how the major sections contribute to the whole and to an understanding of the topic.
- **CCSS.ELA-Literacy.WHST.6-8.2.** Write informative/explanatory texts, including the narration of historical events, scientific procedures/experiments, or technical processes.
- **CCSS.ELA-Literacy.WHST.6-8.2.c.** Use appropriate and varied transitions to create cohesion and clarify the relationships among ideas and concepts.

Background

Hot air rises and cold air sinks, so winds move between the equator and the polar regions. As Earth turns, those winds are twisted by the Coriolis effect and drive surface ocean currents. On a large scale, the work of the wind on the water is predictable. So predictable, in fact, that scientists knew that the Pacific Ocean should have a huge, floating garbage dump even before it was found. The Great Pacific Garbage Patch is a testament to both the movement of Earth's winds and the effect of humans on the Earth. Students will begin by modeling wind movements and then read about the resulting Garbage Patch. They will also examine from a map how the hot and cold currents affect global weather.

Materials

SAFETY NOTE
As you may have observed, push pins are sharp. Don't stick them into your skin.

- Copies of polar projections of Northern and Southern Hemispheres
- Cards reading "hot" and "cold"
- 2 pieces of plastic sheeting (such as overhead sheets)
- Overhead marker
- Corrugated cardboard
- 2 straight edges
- 2 push pins

- Tape
- Wet paper towels
- Globe (optional)
- Pan of water (not round)
- Drinking straws (paper, if possible)
- 3–5 flat pieces of plastic about 1 cm across (cut from an empty container)
- 1 copy of the ocean currents map
- 1 copy of the global wind patterns on plastic sheeting

Student Pages

- Ocean in Motion (lab sheet)
- Polar Projection Maps and Global Wind Patterns (needed to prepare stations)
- "The Garbage Collectors" (article)
- Currents and Climate (thinking visually)

Exploration/Pre-Reading

In this activity, students will explore how wind and surface ocean currents are related. They will observe the Coriolis effect, see how the curving winds push ocean surface waters, and observe the gyres that develop as a result. Students can move through the three stations in any order. If you are concerned that the student will not read the instructions carefully, assign roles with one student as the reader.

To set up Station 1, you will need to make two turntables, as shown in Figure 11.1. Stack two circular pieces of corrugated cardboard, a polar projection map of the Northern Hemisphere, a large circle of clear plastic, and the straight edge (such as a piece of card stock). Stick a pin through the center. Then tape the straight edge to the table so that it won't spin when the rest of the assembly turns. Repeat with the Southern Hemisphere. Label the Northern Hemisphere with an arrow that shows counterclockwise movement, and label the Southern Hemisphere with an arrow showing clockwise movement. Once these are made, you can save them for

Figure 11.1. Station 1

reuse. You may want to place a globe at this station, because it can be difficult for students to visualize why the two hemispheres should be turned in opposite directions without watching a globe turn from the top and the bottom. This station will also need the "hot" and "cold" cards, an overhead pen, and a wet paper towel to clean the plastic between groups.

At Station 2, place the pan of water, drinking straws, and plastic pieces.

For Station 3, copy the wind patterns onto a clear plastic sheet (such as an overhead sheet) and copy the map onto paper or card stock. If you adjust the size of one of the copies, make sure to adjust the other as well, so that they match up when stacked.

Introduce the Reading. Tell students they are going to read an article that talks about one of the consequences of the global surface currents they have been investigating.

Reading Strategy: Identifying Text Signals for Cause and Effect

Begin by displaying the following excerpt from the reading:

> In 1997, Charles Moore was heading home from a sailboat race. His team was in good spirits because …

Ask students to predict what will be at the end of this sentence. Students usually guess that the team won, which is a good guess. But point out that at the very least, you can be pretty sure that the end of the sentence will tell *why* the team was in good spirits. Ask what word signaled what was coming (because). Put up the end of the sentence to confirm their predictions:

> … they had won third place, and they had some extra fuel.

Tell students that certain words and phrases are signals for what the text is about to tell you. *Because* is a signal word for cause and effect. It's especially important in science writing to be on the lookout for words that signal cause and effect because so much of science is the business of figuring out what causes what.

Put the next sentence on the board:

> So they decided to take a different route home, through a part of the ocean that is usually avoided: the North Pacific Subtropical Gyre.

Tell students that this sentence describes an effect. Ask, "What caused the decision to take a different way home?" (They were in good spirits and had extra fuel.) Then ask, "In this sentence, what word signaled that there was a cause and effect?" (so).

Now display the end of the paragraph:

Most ships stay out of the gyre because the wind in the center is weak, and it isn't a good place to catch fish.

Ask what the signal word is (*because*), and explain that as you notice cause and effect while reading, you can use the kind of graphic organizer shown in Figure 11.2 to keep up with the connections:

Figure 11.2. Cause and Effect Graphic Organizer

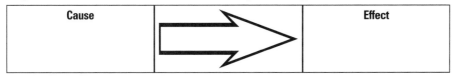

Ask what other words might signal cause and effect. If students have difficulty, add the words *cause / are caused by, consequently, as a result, therefore, for this reason, thus, hence, in response to,* and *since* to the list.

Have students watch for cause-and-effect text signals and underline them as they read. Alternately, underline these text signals yourself and tell students to pause at each one to look for the cause and its effect.

Journal Question

Think about something that happened to you today that had a cause and effect. Write a sentence or two describing it. Use a text signal to help your reader see the connection between the cause and effect.

Application/Post-Reading

- Thinking Visually: Currents and Climate
- Writing Prompt: Cargo ships sometimes lose all or part of their load during storms at sea. Suppose a ship loaded with rubber duckies spilled its cargo off the coast of Morocco in North Africa. Describe two locations, other than the coast of Africa, where the ducks might end up and tell why they might end up there.
  - Prewriting Questions: Help students locate Morocco on a map showing ocean currents and ask (1) What science terms will you

FIND OUT MORE

To learn more about ocean currents and trash circulation, see

- Burns, L. G. 2007. *Tracking trash: Flotsam, jetsam, and the science of ocean motion.* Boston: Houghton Mifflin.

want to include in your writing? (2) You may want to use writing words related to cause and effect. What kinds of writing words would be helpful? (*because, therefore, so,* etc.)

- Key Evaluation Point: Any locations along the currents flowing south and then west would be valid responses. Two likely places would be along the east coast of the United States and in the garbage dump found in the North Atlantic gyre.

Ocean in Motion

In this series of activities, you will explore the movement of wind and water around the world.

Station 1: Wind on a Spinning Earth

1. Look at the polar projection of the Northern Hemisphere. This map shows the Earth as if you were looking down on it from the top. Lines of latitude are shown in circles moving outward. Which circle shows the coldest part—the circle in the middle or the circle on the outside? _____ (Place the card that says "cold" on this circle.)

2. The outermost circle is the equator. Place the cards that say "hot" around the equator.

3. Remember that air molecules move from where they are densely packed to where they are less dense. Which is denser, cold air or warm air?

4. So will the wind tend to blow toward the North Pole or toward the equator?

5. Remove the "hot" and "cold" cards from your map. Then, using the marker provided, trace a straight line from the Pole to the equator.

6. The Earth doesn't sit still while the wind blows. It is turning on its axis. Looking down on the Northern Hemisphere, it turns counterclockwise. Draw the wind blowing in a straight line again, but this time, have a friend turn the map counterclockwise as you trace a path from the Pole to the equator. What happens to the line?

7. Looking up at the Southern Hemisphere, the Earth turns clockwise. Draw a line from the South Pole to the equator while a friend turns the map clockwise. How is this line different from the one on the map of the Northern Hemisphere?

Please gently wipe the lines off of the plastic before moving to your next station.

Station 2: Wind and Water

1. Blow gently through the straw across the surface of the water. What happens to the water?

2. What happens to the direction of the water when it hits the edge of the container?

3. You are modeling the effect of wind on ocean movement. On the basis of this model, how would you expect wind to affect water on the ocean's surface?

4. What might happen when the water encounters a continent?

5. Place the plastic bits in the center of your container. Blow across the surface of the water near one of the edges. Blow as steadily as you can for 20–30 seconds. Describe the movement of the plastic bits.

Safety Note: When you are finished with your straw, tie it in a knot and throw it away. This will ensure that no one accidently reuses your straw.

Please remove the plastic bits and wipe up any water that has splashed before you leave this station.

Station 3: Winds and Currents

1. Look at the map of major ocean surface currents. Locate the United States and look at the currents flowing in the Atlantic Ocean between the United States and Africa. What general shape do the currents make?

2. These currents form the North Atlantic Gyre. Does the North Atlantic Gyre flow clockwise or counterclockwise?

3. Now locate South America. Look at the currents between the east coast of South America and the southern portion of Africa. This is the South Atlantic Gyre. Are the currents flowing clockwise or counterclockwise?

4. Take the diagram of global wind patterns and lay it on top of the map. Now you can see wind patterns and ocean currents at the same time. Compare the wind and water patterns for the North Atlantic Gyre and the South Atlantic Gyre. What do you notice?

Part 4: Final Questions

1. Using information from the stations, what causes wind paths that would normally be straight to curve?

2. At Station 3, you observed the direction of motion for the North and South Atlantic Gyres. Use the information from all of these stations to explain the shape and direction of those two gyres.

Polar Projection Maps and Global Wind Patterns

Northern Hemisphere

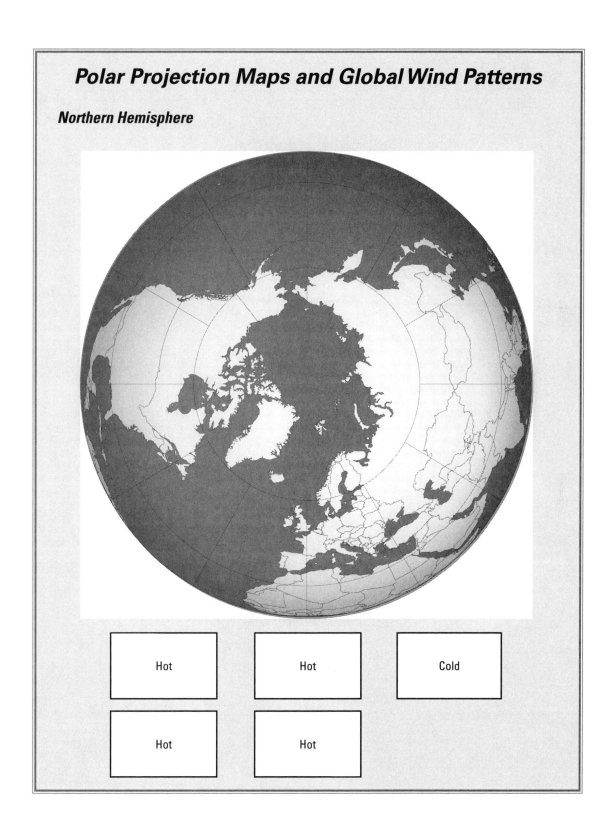

Hot	Hot	Cold

Hot	Hot

Southern Hemisphere

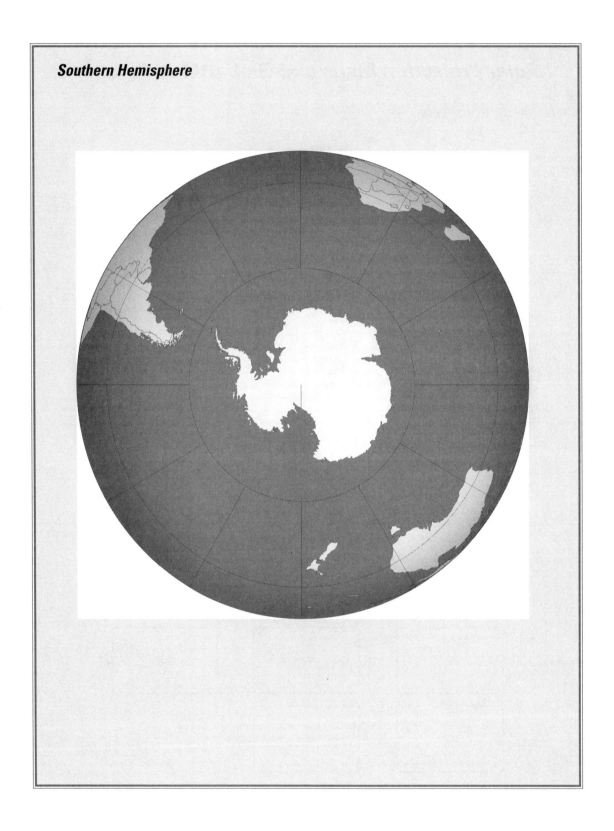

Global Wind Patterns (Excluding Polar Winds)

Major Ocean Surface Currents

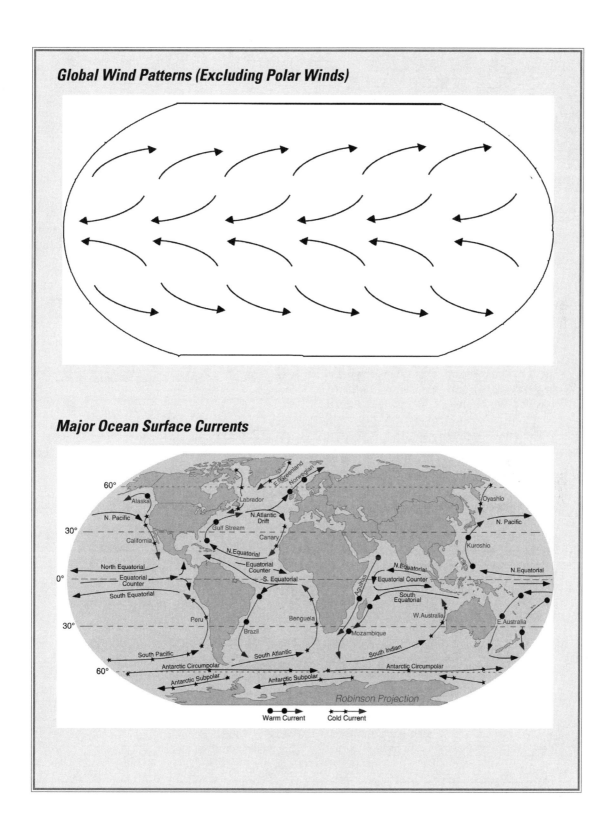

The Garbage Collectors

In 1997, Charles Moore was heading home from a sailboat race. His team was in good spirits because they had won third place, and they had some extra fuel. So they decided to take a different route home, through a part of the ocean that is usually avoided: the North Pacific Subtropical Gyre. Most ships stay out of the gyre because the wind in the center is weak, and it isn't a good place to catch fish.

As he traveled through the gyre, looking out on the ocean, what Charles saw surprised him. His ship was alone in a barely traveled bit of ocean, yet there was evidence of people everywhere. As far as he could see, the ocean was littered with trash.

When he got home, he shared what he had seen. But not everyone was surprised by the discovery. In 1988, a group of scientists at the National Oceanographic and Atmospheric Administration had predicted that trash would be accumulating in this area. These scientists had sampled ocean water for trash and then used their knowledge of ocean currents and winds to determine where the trash was most likely to gather.

Why Here?

The large surface currents in the ocean are caused by wind. In the polar regions at the top and bottom of the Earth, the air is cold. Molecules of cold air pack together and are dense. Near the equator, the Sun's heat warms the air. The molecules spread out and form a low-pressure zone. The polar air rushes toward the equator, where it warms and completes the loop. In addition to moving north and south on the globe, air currents also circulate higher and lower in the atmosphere. As a result, there are a series of circulating air pockets that move heat around the world, as shown in Figure S11.1.

Figure S11.1. Winds and Atmospheric Circulation

Wind moves up and down in the atmosphere and across the globe

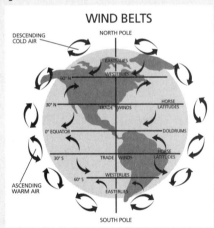

However, Earth's rotation complicates the wind patterns. Because Earth is turning on its axis, the path of the wind, as it is traced across the land, seems to bend. It turns clockwise relative to the Northern Hemisphere and counterclockwise relative to the Southern Hemisphere. This twisting is called the Coriolis effect.

The curved winds then blow across oceans, pushing the surface waters into curved currents. In ocean basins, these currents form loops, or gyres, with relatively calm water trapped in the middle. Trash that travels these currents eventually reaches the center and stays, as shown in Figure S11.2. Scientists have now confirmed the existence of garbage patches in each of the major ocean gyres.

Figure S11.2. Location of Trash Patches in the Pacific

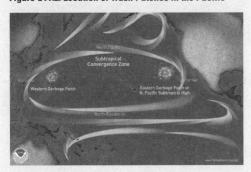

The Plastic Problem

Most of the trash in the garbage patches is plastic. Unlike paper or food scraps, plastic never breaks down into material that living things can reuse. It just breaks into smaller and smaller pieces. For this reason, the garbage patches don't look like a sea of plastic cups, detergent bottles, or potato chip bags. By the time the plastic makes it to the garbage patches, it has broken into chips and bits that look more like ingredients in a plastic soup, as shown in Figure S11.3. Therefore, cleanup is almost impossible. The plastic bits are as small as many of the living things in the ocean. A scoop that could remove the plastic would also capture the living things that are the base of ocean food chains.

Figure S11.3. Plastic Soup From an Ocean Gyre

Unfortunately, animals don't seem to be able to sort out what is plastic and what is living, either. Both fish and seabirds have been found with bellies full of plastic bits. Figure S11.4 shows the body of an albatross, a large seabird, which had plastic in its digestive system when it died.

Figure S11.4. Signs That Animals Eat Plastic

The body of a seabird whose digestive system was filled with plastic

There is currently no plan for cleaning up the garbage patch. The best solution would be to use less plastic, so that less of it falls into the ocean, catches the surface current, and finds its way into the ocean gyres.

THE BIG QUESTION

Explain how wind and currents cause garbage to collect in specific places on Earth.

Currents and Climate

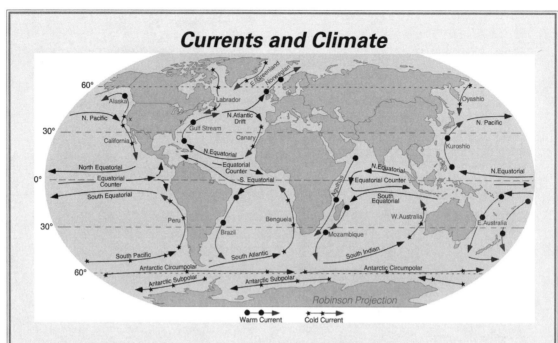

Robinson Projection

● ● ● ➤ Warm Current   ✴—✴—✴➤ Cold Current

1. Scan the map above for cold water currents. In what general area are most cold water currents found?
2. In what general area are most warm water currents found?
3. Why would warm currents originate near the equator?
4. Look for one of the places where a current brings cool water toward the equator. Describe the current's path.
5. Look for one of the places where a current brings warm water away from the equator. Describe the current's path.
6. How might those currents affect the temperature on land nearby?
7. On the map, San Francisco, California, is marked with an X. Hampton, Virginia, is marked with a Y. Both of these cities are at about 37° latitude. Therefore, they receive about the same amount of sunlight during the year. Look at the graph of their high temperatures below. Use the map to explain why these two cities have such different temperatures.
   A. Which city has the cooler temperature in the summer?
   B. What could explain that city's cooler temperature?

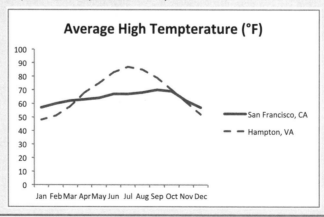

Average High Tempterature (°F)

San Francisco, CA

Hampton, VA

NATIONAL SCIENCE TEACHERS ASSOCIATION

Fury in the Water

Topics
- Specific heat of water
- Wind
- Hurricanes

Reading Strategy
- Identifying text signals for comparisons and contrasts

Lesson Objectives: Connecting to National Standards

The following list shows the *Next Generation Science Standards* (*NGSS*) and *Common Core State Standards* (*CCSS*) supported by this activity.

NGSS: *Science and Engineering Practice*
- Engaging in Argument From Evidence

NGSS: *Disciplinary Core Idea*
- **ESS2.D.** Weather and Climate

NGSS: *Crosscutting Concept*
- Systems and System Models

CCSS: *Literacy in Science and Technical Subjects*
- **CCSS.ELA-Literacy.RST.6-8.1.** Cite specific textual evidence to support analysis of science and technical texts.
- **CCSS.ELA-Literacy.RST.6-8.9.** Compare and contrast the information gained from experiments, simulations, video, or multimedia sources with that gained from reading a text on the same topic.
- **CCSS.ELA-Literacy.WHST.6-8.1.** Write arguments focused on discipline-specific content.

Background

FIND OUT MORE
To learn more about hurricanes, visit the National Hurricane Center at *www.nhc.noaa.gov.*

The Sun is the ultimate power source for many of the processes on Earth, and water serves as a battery, storing and releasing the Sun's power. These ideas are important for students to understand, because they form the basis of weather. In this chapter, students will discover how the power of the Sun and the heat capacity of water produce wind and hurricanes.

Materials

SAFETY NOTE
Heat lamp and incandescent bulbs get extremely hot. Students who have spent their lives around LED and CFL bulbs may not be aware of how hot lightbulbs can get. Be sure to warn them not to touch the bulbs.

For the first demonstration:
- 2 identical lamps with incandescent or heat lamp bulbs
- Identical containers that hold at least $1/3$ cup
- 2 mercury-free thermometers
- $1/3$ cup sand
- Water

For the second demonstration:
- Candle in a candle stand
- Lighter or matches
- 2 balloons
- Funnel
- Water
- Sanitized safety glasses or goggles

Student Pages
- Pop! Lab (lab sheet)
- "The Fury in the Water" (article)
- Breezy Beaches (thinking visually)

Exploration/Pre-Reading

This lab has two parts. The first activity models Sun shining on the beach and the ocean. For best results, allow the lamps to heat the water and sand for at least 40 minutes, which means it is preferable to do this portion of the lab at the beginning and ending of class the day before doing the second half of the lab. Alternatively, the students can take the starting temperature before doing the group activity and then complete the first section of the lab about 20 minutes later. With the shorter time period, it is best to use a heat lamp or locate a 100 W incandescent bulb. On the lab sheet, this activity is described as a demonstration because it can be expensive to have enough heat lamps for the class. However, it works well in small groups if you have the equipment.

For the second demonstration, blow up a balloon and tie it off. Set a candle on a lab table and hold the balloon just above—but not touching—the flame. Have students count to 10 while you hold the balloon in the heat. It will pop, usually long before 10. Next, fill a second balloon with as much water as it will hold without expanding. Then blow it up the rest of the way with air and tie it off. Repeat the process of holding it over (but not in!) the flame as students count to 10. You can really ham this up by acting like you think the water balloon is going to pop and make a mess. Students will be surprised when the second balloon does not pop. It is best to use new, high-quality balloons because older and cheaper balloons are more likely to pop unexpectedly. Have students summarize what they observed on their data sheet.

Introduce the Reading. Tell students they are going to read an article that will explain some ways that the high heat capacity of water affects things that happen on Earth.

Reading Strategy: Recognizing Signal Words for Compare and Contrast

Begin by displaying the following sentence from the reading, with blanks as shown:

In the same way that sand heats up quickly, it cools _____. The air over the sand also cools.

Remind students that certain words are signals for what the text is about to say. (Text signals are first introduced in Chapter 11, p. 109. If you

SAFETY NOTE
Be sure to wipe up any spilled water to prevent slip or fall hazards. Keep water away from electrical equipment to prevent shock. Only use GFI-protected electrical circuits for lamps. Wash hands with soap and water upon completing the activity.

SAFETY NOTE
This demonstration will need to be conducted with latex balloons. Do not perform it if you have a latex allergy. Dispose of latex balloon parts carefully, and wash your hands with soap and water to protect students who may have a latex allergy.

SAFETY NOTE
Model good safety practices. Wear sanitized safety glasses or goggles meeting the ANSI Z87.1 standard during the setup, demonstration, and takedown.

did not use Chapter 11 with your class, you may need to provide more background on text signals than is given here.) See if anyone can predict what word might go in the blank. If they need a hint, underline the signal words *in the same way*. Otherwise, ask which words helped them fill in the blank. *In the same way* is a signal that two things are being compared. When they come across signal words for comparisons, they should ask themselves two questions: (1) What things are being compared? And (2) how are they alike? Ask students to list other words that might signal a comparison (*likewise, just like, just as, also, too*).

Now add this sentence beneath the previous sentence, with blanks as shown:

> *In contrast,* water heats _____, but it *also* cools _____.

Ask students to predict what might go in the blank (slowly), and point out that *in contrast* indicated that water would be different from sand. *In contrast* signals that the new information is going to be different from, or contrast with, earlier information. When students come across contrast signal words, they should ask themselves what things are being contrasted, and how they are different. Ask students what other words might signal a contrast (*however, on the other hand, conversely, whereas, but, yet,* and *while* sometimes indicate a contrast).

Journal Question

Imagine that a friend asks you what a signal word is. What would you tell your friend? What signal words would you recommend they look for?

Application/Post-Reading

- Thinking Visually: Breezy Beaches
- Writing Prompt: Think back to your lab. Why didn't the balloon pop when it was filled with water? Make a claim, support it with evidence, and explain how the evidence supports your claim.
  - Prewriting Questions: Make a quick list of the information that you want to include in your explanation. Number the ideas in the order that you think you will introduce them as you write. What science words will you want to include?

○ Key Evaluation Point: Water has a high heat capacity—much higher than that of air. When the first balloon was heated, the temperature of the air increased quickly, and the rubber melted. When the second balloon was heated, the water's high heat capacity meant that the water's temperature increased more slowly. The rubber did not get hot enough to melt.

Pop! Lab

Part 1: *In Hot Water*

Set up two identical desk lamps and place identical containers beneath each one. Fill one container with ¹/₃ cup of water and the other with ¹/₃ cup of sand. Record the starting temperatures. Allow the sand and water to sit under the light for at least 40 minutes, and then record their temperature.

1. The containers are getting equal amounts of heat from the lights. What do you predict will happen?

2. Complete the data chart:

Temperature	Sand	Water
Starting Temperature		
Ending Temperature		

Part 2: *Pop!*

Your teacher will perform a demonstration. Sketch the setup and record the results below:

Part 3: *Heat Capacity*

Heat capacity is a measure of how much heat it takes to warm a substance up. Here are the heat capacities of water, air, and sand:

Substance	Joules of Energy Needed to Warm 1 kg by 1°C
Air, Dry (Sea Level)	1,005
Water, Pure	4,186
Quartz Sand	830

1. According to the chart, which substance needs more energy to get hot: sand or water?

2. In Part 1 of this experiment, which substance needed more energy to get hot: sand or water?

3. According to the chart, which substance needs more energy to get hot: air or water?

4. With the first balloon in Part 2, rubber and air were absorbing the heat from the candle. The rubber quickly melted. What was absorbing the heat from the candle in the second balloon?

Part 4: *Make a Claim*

Why didn't the balloon pop when it was filled with water? Make a claim, support it with evidence, and explain how the evidence supports your claim.

The Fury in the Water

Picture a hot day at the beach. People squint their eyes against the light from the Sun and jump in the water to escape the Sun's heat. Both light and heat are forms of energy. That energy affects land and water just as it affects people.

The sand heats up first. It doesn't take a lot of energy to heat sand. Heating 1 kg of sand 1°C requires about the same amount of energy that it takes to light a lightbulb for 26 seconds. In contrast, it takes about five times that much energy to warm 1 kg of water. Water has a high heat capacity, which means that it takes a lot of energy to get it hot.

As the day goes on, the sand gets much hotter than the ocean next to it. The hot sand warms the air around it, and the warm air rises. In contrast, the air over the cooler ocean water isn't as hot. When the warm air over the sand rises, the cooler air near the water blows into the spaces left behind. By lunchtime, a strong wind is moving from ocean to land, as shown in Figure S12.1.

Figure S12.1. Movement of Air During the Day

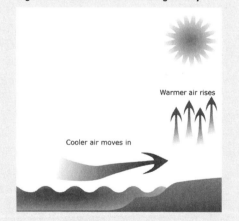

The Sun sets in the evening and stops pouring energy over the land and sea. In the same way that sand heats up quickly, it cools quickly. The air over the sand also cools.

In contrast, water heats slowly, but it also cools slowly. As night falls, the ocean is still releasing heat into the air above the water. Now the air on land is cooler, and rushes into the spaces left by the warm, rising ocean air. The wind blows from shore to sea, as shown in Figure S12.2.

Figure S12.2. Movement of Air During the Night

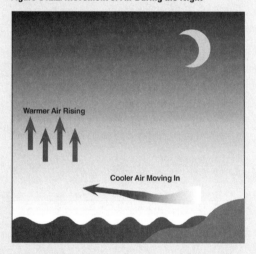

Therefore, at the coast, water's high heat capacity causes winds to blow to and from the shore. In other places, however, that same high heat capacity turns violent.

Heat-Powered Hurricanes

Near the equator, the Sun beats down on the ocean water for long hours every day, all year long. The water absorbs much more heat energy than it releases during the night. The water gets hotter. The energy builds up day after day. And each day, the water heats the air just above it, and the hot air rises. Then cooler air rushes in.

Most of the time, this pattern is fairly calm, and helps move energy from the equator to cooler areas of the Earth. But when the ocean is especially hot, a weather pattern can develop that concentrates this upward motion in a small area, as shown in Figure S12.3. However, it's not just air that's moving up. Massive quantities of water droplets evaporate from the surface of the ocean and join the swirling air, creating drenching rainfall. And what about the energy that's stored in those water droplets? When the water condenses and falls as rain, it releases much of that energy back into the storm.

Figure S12.3. The Beginnings of a Hurricane

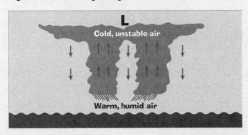

Hot air rises, cool air moves in. Water vapor condenses, releasing its energy into the air. More hot air rises, more cool air moves in. ... Soon swirling clouds develop. The spinning Earth pushes the clouds into a spiral. When the winds reach 39 mph, we call it a tropical storm. When they reach 74 mph, we call it a tropical cyclone, a typhoon, or a hurricane.

Weather to Watch

Hurricanes and tropical storms start in the warmest waters of the ocean, but global winds move them north or south away from the equator. They can be huge. Hurricane Sandy, in 2012, was almost 1,000 miles wide at its largest and killed 149 people. These storms can last for weeks. In 1971, Hurricane Ginger lasted almost an entire month. Therefore, one storm can affect millions of people.

Fortunately, satellites, radar, and other tools can detect tropical storms as they are forming over the ocean (Figure S12.4). Hurricane forecasters gather information about the storms as they develop. Then they use our knowledge of global weather patterns to predict the most likely paths that a hurricane may take. This allows people in the path of the hurricane to find a safe place to ride out the storm.

Figure S12.4. Satellite Image of a Hurricane

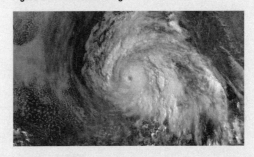

A gentle breeze at the beach and a violent storm over the ocean don't at first seem to have much in common. Yet, they are driven by the same engine: the high heat capacity of water. Wherever water absorbs and stores energy, it affects the weather nearby. And that weather, whether fierce storms or gentle breezes, redistributes the energy of the Sun around the Earth.

THE BIG QUESTION

How does water's heat capacity cause the wind to blow at the beach?

Breezy Beaches

1. Sometimes, the best way to understand information is to think it through in a picture. Write each of the phrases below on either the water side or the land side of the picture of a beach during the day.

   - higher heat capacity
   - lower heat capacity
   - heats quickly
   - heats slowly
   - warmer air above
   - cooler air above

2. Next, identify which side of the picture has air that is *less dense.* Draw an arrow showing that the air on that side of the diagram rises.

3. Draw a second arrow to show the direction that the wind will blow in this diagram.

4. Now use the same phrases and arrows to label the picture below with what happens at night.

On the Outside Looking in

Topics
- Solar system size and scale
- Inner and outer planets
- Formation of the solar system

Reading Strategy
- Previewing diagrams and illustrations

Lesson Objectives: Connecting to National Standards

The following list shows the *Next Generation Science Standards* (*NGSS*) and *Common Core State Standards* (*CCSS*) supported by this activity.

NGSS: *Science and Engineering Practices*
- Developing and Using Models
- Analyzing and Interpreting Data

NGSS: *Disciplinary Core Idea*
- **ESS1.B.** Earth and the Solar System

NGSS: *Crosscutting Concepts*
- Patterns
- Scale, Proportion, and Quantity
- Systems and System Models

CCSS: *Literacy in Science and Technical Subjects*
- **CCSS.ELA-Literacy.RST.6-8.7.** Integrate quantitative or technical information expressed in words in a text with a version of that information expressed visually (e.g., in a flowchart, diagram, model, graph, or table).
- **CCSS.ELA-Literacy.RST.6-8.9.** Compare and contrast the information gained from experiments, simulations, video, or multimedia sources with that gained from reading a text on the same topic.
- **CCSS.ELA-Literacy.WHST.6-8.2.** Write informative/explanatory texts, including the narration of historical events, scientific procedures/experiments, or technical processes.

Background

It is hard for anyone to conceive of the sheer size and the vast amount of empty space in our solar system, much less a galaxy or the universe. It is much easier for the planets to be reduced to a list of their properties. In this chapter, we'll fight that tendency by inviting students to explore the size of the solar system and the properties of the planets in a way that should help them understand the larger story of our solar system. They will also learn about NASA's Juno mission and, hopefully, develop an interest in other aspects of space exploration.

As you talk with students about this topic, be aware that they often confuse the solar system, our galaxy, and the universe. A common misconception is that the birth of our solar system is the big bang. Be on the lookout for these misconceptions and address them as they come up.

Materials

- Tape measure
- Masking tape
- Clay
- Styrofoam

- Toothpicks or kabob sticks
- Basketball
- 8 glass or clear plastic jars
- Marbles or dried beans

Student Pages

- Planet Cards (lab material)
- Student Instructions for Each Station (lab material)
- Exploring the Solar System (lab sheet)
- "To Jupiter, Juno!" (article)
- Planet Properties (thinking visually)

Exploration/Pre-Reading

Before class, gather the materials and set up each of the five stations. With large classes, you may prefer to have two of each station so that students can work in smaller groups. To ensure that students read the background information at each station, you may wish to assign roles that include a reader.

To set up, place a copy of the student instructions for that station, along with the materials described below, at each station.

Station 1: How Far Apart Are the Planets?

Before class, mark the scaled distances of the planets along a hallway or outside. Table 13.1 (p. 136) provides a scale that would work in a 50 m hallway. However, you can use the chart to make a scale that works for whatever size hallway you have. To make measuring easier, you may want to scale the distances based on the sizes of the tiles in your hallway or the length of your stride. Write your scale in the blanks on the student station instructions. Students will walk the distances at this station using "heel-toe" walking to help them experience the large distances kinesthetically.

> **TEACHING NOTE**
> There are many activities for exploring the distance between planets. Feel free to adapt any that you particularly like. To create this station, I used ideas from Davies, M., L. Landis, and A. Landis. 2009. Solar system in the hallway. *Science Scope* 33 (April–May): 56–60.

Table 13.1. Planet Orbits Scaled to a 50 m Hallway

Body	Average Orbit Radius (km)	Scaled Orbit (m) (1 m = 90,000,000 km)
Sun	0	—
Mercury	58,000,000	0.64
Venus	108,000,000	1.2
Earth	150,000,000	1.7
Mars	228,000,000	2.5
Jupiter	778,000,000	8.6
Saturn	1,430,000,000	15.9
Uranus	2,870,000,000	31.9
Neptune	4,490,000,000	49.9

Station 2: How Big Are the Planets?

Before class, prepare clay models of each planet according to the scale in Table 13.2. Stick each planet on a toothpick or kabob stick and mount them on a Styrofoam block, as shown in Figure 13.1. Although a bit small, a standard basketball is a reasonable representation of the Sun on this scale. At the station, students will view the scale models and a chart of the diameters of the planets. You may also wish to provide a printout of scaled drawings of the planets; NASA has an excellent one at *http://solarsystem. nasa.gov/multimedia/gallery/solarsys_scale.jpg.*

This may be the hardest station for students to grasp because the concept of scale is difficult for many students. You may need to interact with the students at this station to ensure they understand what is being said.

SAFETY NOTE
Review the Safety Data Sheet for clay. Do not allow clay to dry and form dust on furniture or floor due to potential silica exposure. Wash hands with soap and water on completing the activity. Don't poke yourself with the toothpicks or kabob sticks.

Table 13.2. Planet Scale for Station 2

Body	Approximate Diameter (km)	Scale (mm) (1 mm = 5,000 km)
Sun	1,392,000	278
Mercury	4,980	1
Venus	12,360	2.5
Earth	12,742	2.5
Mars	6,760	1.4
Jupiter	142,600	28.5
Saturn	120,600	24.1
Uranus	47,000	9.4
Neptune	44,600	8.9

Figure 13.1. Planet Models Mounted on a Styrofoam Block

Station 3: Moons and Rings

Before class, fill a jar with marbles or beans representing the number of moons for each planet. Label the top of the jars as shown in Table 13.3.

Table 13.3. Jar Labels for Station 3

Planet	Label
Mercury	0
Venus	0
Earth	1
Mars	2
Jupiter	Up to 67 (17 objects may be moons, but more data are needed to be sure)
Saturn	Up to 62 (9 objects may be moons)
Uranus	27
Neptune	Up to 14 (1 object may be a moon)

Station 4: What Are the Planets Made Of?

Make a copy of the Planet Cards that show the composition of the planets. Cut the planets apart so that students can interact with them individually and move them around to compare them.

Introduce the Reading. Tell students that they are going to read an article about a NASA space mission. Have them watch for information that

> **TEACHING NOTE**
> NASA has an excellent website for the Juno mission (as it does for most missions). The website includes a number of videos that are two minutes or shorter. Once Juno has reached Jupiter in 2016, there will even be ways that amateurs can help process the data coming in. There are many possible extensions to this chapter at *http://missionjuno.swri.edu.*

would explain why the same planets kept ending up in the same groups during their exploration.

Reading Strategy

Tell students that in some books they read, the pictures may be extras. In science writing, however, the pictures and diagrams often carry a lot of important information. Looking at the pictures and making predictions about what they mean before reading can help make the text easier to understand.

Place students in reading groups and hand out "To Jupiter, Juno!" but tell students not to read yet. Guide students through looking at the diagrams using the following prompts. For the prediction questions, accept any answer as a valid possibility, and tell students they will have to read the text to find out if they are correct.

Start with Figure S13.1 (p. 148). Ask each group member to answer a question in their group. Then have one group share with the class. (Note that the role titles are only used to ensure that each group member participates in the discussion.)

Leader: What do you think this object is? (prediction)

Interpreter: Where is it located and what clues from the picture let you know?

Direct them to Figure S13.2 (p. 148).

Emergency Manager: Describe what you see in this picture. Don't worry if you don't know some of the names of things. Just describe what you see.

Leader: What do you think the thermometer is there for? (prediction)

Interpreter: What could the frost line be? (prediction)

Finally, look at Figure Figure S13.3 (p. 149).

Emergency Manager: Read the caption for this picture. What does it mean that "the sizes are to scale?"

Leader: Why aren't the distances to scale? (*Hint:* think about our exploration.)

Interpreter: Remember the frost line from Figure S13.2? Take a guess where the frost line would fall in this diagram. (prediction)

Journal Question

Do you typically study the diagrams and pictures in your science book before you read? Was it helpful to look at the diagrams and pictures before reading today? Why or why not?

Application/Post-Reading

- Thinking Visually: Planet Properties
- Writing Prompt: In another solar system in our galaxy, scientists have found a planet that is a lot like Jupiter. It is large, not very dense, and mostly made of light gases. However, this gas giant orbits very close to its star. Scientists suspect this gas giant's orbit may have gotten smaller at some point. Assuming that solar system developed the same way that ours did, why would scientists suspect that the planet has moved?

  FIND OUT MORE
  To learn more about the birth of the solar system, see NASA's Solar System Exploration site at *http://solarsystem.nasa.gov.*

  - Prewriting Questions: (If you are having trouble thinking about this question, consider drawing a diagram of the change in orbit that scientists think happened and comparing it to Figure S13.2 in the article.) (1) What science words should you include in your response? (2) You will probably do some comparing and contrasting as you write. What writing words might you want to use? (Suggest words from Table 2.2, p. 18, in Chapter 2 to students.)
  - Key Evaluation Point: Because solar wind pushes the lighter elements away after a star has formed, large gaseous planets usually form far away from the star. If this planet looks like it should be far away, but it isn't, it may have moved.

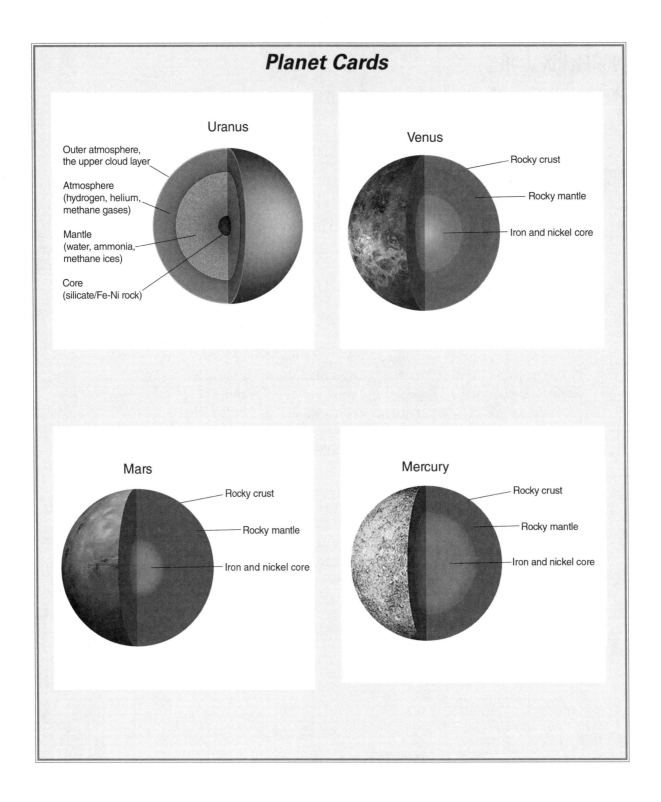

Planet Cards

Uranus

Outer atmosphere, the upper cloud layer

Atmosphere (hydrogen, helium, methane gases)

Mantle (water, ammonia, methane ices)

Core (silicate/Fe-Ni rock)

Venus

Rocky crust

Rocky mantle

Iron and nickel core

Mars

Rocky crust

Rocky mantle

Iron and nickel core

Mercury

Rocky crust

Rocky mantle

Iron and nickel core

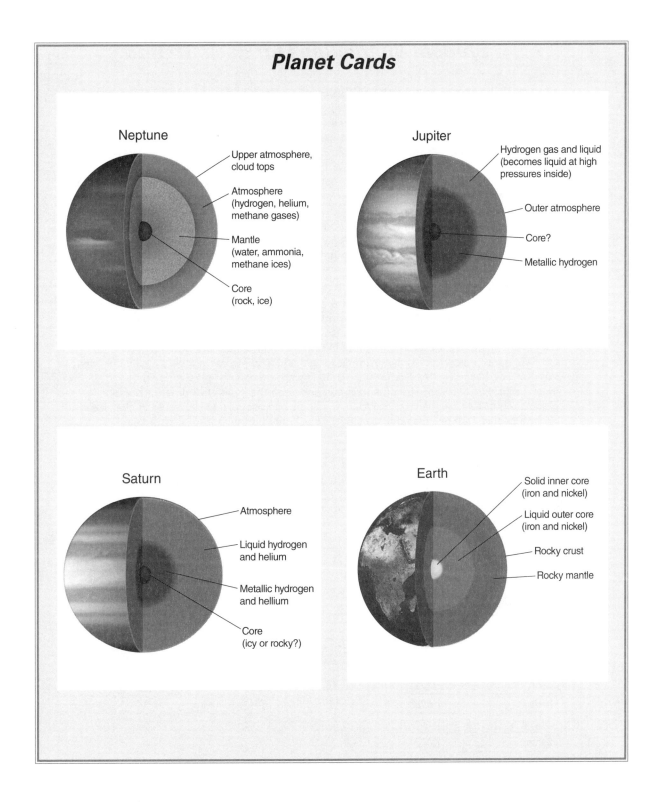

Planet Cards

Neptune
- Upper atmosphere, cloud tops
- Atmosphere (hydrogen, helium, methane gases)
- Mantle (water, ammonia, methane ices)
- Core (rock, ice)

Jupiter
- Hydrogen gas and liquid (becomes liquid at high pressures inside)
- Outer atmosphere
- Core?
- Metallic hydrogen

Saturn
- Atmosphere
- Liquid hydrogen and helium
- Metallic hydrogen and hellium
- Core (icy or rocky?)

Earth
- Solid inner core (iron and nickel)
- Liquid outer core (iron and nickel)
- Rocky crust
- Rocky mantle

Student Instructions for Each Station

Station 1: How Far Apart Are the Planets?

At this station, you will take a journey across the solar system and visit each planet. The planets are shown here in a *scale model*. In real life, of course, the planets are much farther apart. But we've shrunk the distance between planets in a way that is proportional. Your teacher has set this model up so that _____ = _____ of distance.

1. Measure your walk across the solar system by counting your footsteps. Each time you step, put the heel of your foot directly in front of the toes of the foot that is already on the ground. Count the number of steps between each planet and record them on your data sheet.
2. Complete this sentence: I was surprised that _____.
3. Looking at your data and the planets in the hallway, decide how you could group the planets into two groups based on how far they are from the Sun.

Station 2: How Big Are the Planets?

This station shows a scale model of the Sun and the planets. A scale model shows how big things are in relation to each other, not how big they are in real life. In real life, Earth is two and half times bigger than Mercury. So in this scale model, Earth is two and a half times bigger than Mercury. Of course, the real Earth and real Mercury are much bigger.

It is hard to show the scale of the solar system. The planets are very small compared to how far apart they are. At Station 1, the distance between the planets is shown in the hallway. To fit the solar system in the hallway, and keep the scale the same for the planets, the planets would be tiny. Mercury and Mars would be microscopic. Earth would be the size of the tip of a needle. Jupiter, the largest planet, would only be the size of a small flea.

However, we could set up the solar system using the scale shown here, so that we can at least see Mercury. In that case, Jupiter would have to be more than half a mile away from the school!

1. There are no drawings that show both the distances between planets and the size of the planets accurately to scale. Why not?
2. Table S13.1 shows the actual diameters (distance across the middle) of the planets in kilometers. Looking at this chart and the scale model of the planets, decide how you could sort the planets into two groups based on their size.

Table S13.1. The Diameter of Each Planet

Body	Approximate Diameter (km)
Sun	1,392,000
Mercury	4,980
Venus	12,360
Earth	12,742
Mars	6,760
Jupiter	142,600
Saturn	120,600
Uranus	47,000
Neptune	44,600

Station 3: Moons and Rings

A moon is a planetlike object that orbits a planet instead of a star. The jars here show how many moons are thought to orbit each planet. As technology gets better, scientists are discovering more and more moons!

1. Look at the number of moons. Complete the sentence: It surprised me that _____ _____ .

Rings are a collection of dust, rocks, and ice that orbit a planet (Table S13.2). Some of the particles in a ring are microscopic. Others can be as big as a house.

Table S13.2. The Rings of Each Planet

Planet	Number of Rings*
Mercury	0
Venus	0
Earth	0
Mars	0
Jupiter	1
Saturn	7
Uranus	13
Neptune	6

Source: Data from the 2013 Solar System Lithograph from NASA.
* Known rings as of July 2013.

2. Are planets more likely to have a ring if they are close to the Sun or far from the Sun?

3. Use the number of moons and rings to decide how you could sort the planets into two groups based on the objects that orbit them.

Station 4: What Are the Planets Made Of?

Look at the planet cards to see our current understanding of what the planets are made of. Note that the planets are not to scale in these diagrams!

1. Look carefully at the composition of each planet. Find a way to sort the planets into two similar groups.

Table S13.3 describes the atmosphere of each planet.

Table S13.3. The Atmosphere of Each Planet

Planet	Main Compounds in the Atmosphere
Mercury	no real atmosphere
Venus	carbon dioxide and sulfuric acid
Earth	nitrogen and oxygen
Mars	carbon dioxide (very thin atmosphere)
Jupiter	hydrogen and helium
Saturn	hydrogen and helium
Uranus	hydrogen, helium, methane
Neptune	hydrogen, helium, water, silicates, methane

2. The Sun is mostly composed of hydrogen and helium. Which planets' atmospheres are most similar to the makeup of the Sun?

Exploring the Solar System

Name: _____

Station 1: How Far Apart Are the Planets?
(Read the Station 1 student instructions FIRST!)

Planet	Heel-Toe Steps Away From the Sun

1. I was surprised that _____.

2. Looking at your data and the planets in the hallway, decide how you could group the planets into two groups based on how far they are from the Sun.

Group 1	Group 2

What information helped you decide on this grouping?

Station 2: How Big Are the Planets?
(Read the Station 2 student instructions FIRST!)

1. Why don't drawings show both the size of planets and the distance between planets accurately to scale?

2. Decide how you could group the planets into two groups based on their size.

Group 1	Group 2

What information helped you decide on this grouping?

Station 3: Rings and Moons
(Read the Station 3 student instructions FIRST!)

1. It surprised me that _____.

2. Are planets more likely to have a ring if they are close to the Sun or far from the Sun?

_____

3. Put the planets into two groups based on the objects (moons and rings) that orbit them.

Group 1	Group 2

What information helped you decide on this grouping?

Station 4: What Are the Planets Made Of?
(Read the Station 4 student instructions FIRST!)

1. Place the planets in two groups based on what the planets are made of (ingredients).

Group 1		Group 2	
Four planets	Three common ingredients	Four planets	Three common ingredients

2. Using information from the instruction sheet, which planets' *atmospheres* are most similar to the makeup of the Sun?

Conclusions
Look at the *four* places above where you grouped the planets. Based on *all* of the characteristics you have learned about, decide on a *final* grouping for the planets and describe the characteristics of each group.

Group 1	Group 2
Characteristics of group 1	Characteristics of group 2

To Jupiter, Juno!

In August 2011, a rocket bearing the Juno spacecraft (Figure S13.1) launched from Cape Canaveral, Florida, and raced away from Earth. Juno's destination was Jupiter, the fifth planet from the Sun, and the largest planet in our solar system. If all goes as planned, Juno will arrive in July 2016 and help scientists piece together the story of how our solar system was born.

Figure S13.1. Juno

Juno unfolding its solar panels to capture power for its mission

The Story So Far

Our solar system is made up of the Sun and the eight planets that orbit it. It is just one of billions of solar systems in our galaxy, and our galaxy is one of billions of galaxies in the universe. The universe began a little over 13 billion years ago. Our galaxy was born shortly thereafter. But our solar system is only about 4.6 billion years old.

Our solar system began in a cloud of dust and gas that was probably leftover from a star that had exploded. At first, there was just a spinning disc of debris. But this matter had gravity, and the pieces were attracted to each other. Most of it clumped together in the middle and formed our star, the Sun. But at the same time, smaller clumps were forming in orbit around the Sun. These clumps would eventually become the planets of our solar system.

Once the Sun began to burn brightly, it produced solar wind. The wind blew outward, and pushed the lightest elements toward the outside of the solar system. Heavy metals such as iron and nickel, and the materials that make up rocks, stayed close. Lighter elements such as hydrogen and helium blew toward the edges.

The Sun also created a heat gradient. The parts of the solar system near the Sun were very hot, and as material traveled away from the Sun, it cooled off. Any water that was near the Sun existed only as far-flung molecules of gas. Water could only condense into liquid or ice past the point that scientists call the frost line (Figure S13.2). The first planet formed in this cold, outer region. This first planet, the biggest planet, was Jupiter.

Figure S13.2. Planets Inside and Outside the Frost Line

Different elements were available to planets inside and outside the frost line.

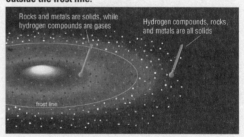

Two Groups

Because Jupiter formed far out in the solar system, it is mostly made of lighter elements. It also has water and other substances that could only condense in colder regions, away from the Sun. It is called a gas giant, and for good reason. It is mostly made of gas. However, the pressure from the outside layers of the planet pushes some of that gas together to form liquid inside. At the center, Jupiter has a core ... maybe. One of Juno's jobs is to find out if there is a core, and if so, what it is made of.

Learning about Jupiter will shed light on other planets as well. The planets form two distinct groups. The outer planets, or gas giants, are those that formed outside the frost line in the early solar system. Saturn, Uranus, and Neptune are similar to Jupiter. They are all large compared with the other planets, but they are not very dense. Much of their volume is made up of gases.

Their gravitational pull is very strong. When the outer planets encounter dust, gases, or rocks, gravity pulls that debris into orbit around the planet. Therefore, the outer planets have rings and many moons.

The inner planets—Mercury, Venus, Earth, and Mars—are smaller. They are rocky, dense, and made of the heavier elements that were not as easily blown by solar wind. They have little or no atmosphere, and few, if any, moons.

This raises one of the most difficult questions in the story of our solar system (Figure S13.3). How did water, which is so crucial to life on Earth, end up on a planet far inside the frost line?

Juno's Job

Figure S13.3. Planets of Our Solar System

The planet sizes are to scale, but the distances between the planets are not.

Juno may reveal some clues to that mystery, too. Data from an earlier mission suggest that Jupiter also has some surprising elements. These elements needed to be even colder than water to condense and should have formed out past Jupiter. Is it possible that some of the planets moved around after they were formed? Or did debris from farther out in the solar system carry these elements inward?

Scientists eagerly await Juno's data from Jupiter, which should settle these questions. Although, perhaps, the data sent back by Juno will report something that no one is expecting. Then scientists will have to reconsider the story of the birth of Jupiter—and of our solar system.

THE BIG QUESTION

Two features of the early solar system caused the inner and outer planets to develop differently. What were they?

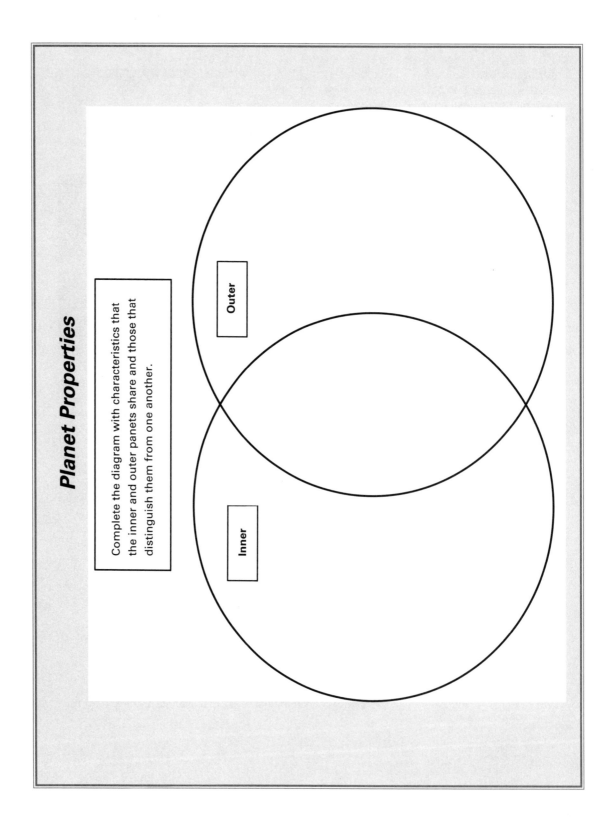

Planet Properties

Complete the diagram with characteristics that the inner and outer panets share and those that distinguish them from one another.

Outer

Inner

The 20-Year Winter

Chapter 14

Topics

- Seasons
- Axial tilt
- Uranus

Reading Strategy

- Previewing diagrams and illustrations

Lesson Objectives: Connecting to National Standards

The following list shows the *Next Generation Science Standards* (*NGSS*) and *Common Core State Standards* (*CCSS*) supported by this activity.

NGSS: *Science and Engineering Practice*
- Developing and Using Models

NGSS: *Disciplinary Core Ideas*
- **ESS1.A.** The Universe and Its Stars
- **ESS1.B.** Earth and the Solar System

NGSS: *Crosscutting Concept*
- Systems and System Models

CCSS: *Literacy in Science and Technical Subjects*

- **CCSS.ELA-Literacy.RST.6-8.7.** Integrate quantitative or technical information expressed in words in a text with a version of that information expressed visually (e.g., in a flowchart, diagram, model, graph, or table).
- **CCSS.ELA-Literacy.WHST.6-8.1.** Write arguments focused on discipline-specific content.
- **CCSS.ELA-Literacy.WHST.6-8.1.a.** Introduce claim(s) about a topic or issue, acknowledge and distinguish the claim(s) from alternate or opposing claims, and organize the reasons and evidence logically.

Background

Many middle school students find the causes of seasons to be a difficult concept. Students often need to work with several visual representations to understand the implications of axial tilt. One particularly stubborn misconception is that seasons change as the Earth moves farther from and closer to the Sun.

In this chapter, students will experiment with using a model to create an explanation for a data set, in this case, the seasonal temperatures of cities north and south of the equator. This helps counter the misconception that seasons are caused by changing distance from the Sun. Under that model, the two cities should have the same seasons at the same time. After creating a solution, students will read more about how seasons are caused both on Earth and on Uranus.

Materials

- Flashlight (can use just two or three for the class)
- 1 3 in. Styrofoam ball (per group)
- Permanent marker
- 2 thin dowels (per group)
- 1 lightbulb that shines in all directions (per group) (see directions below)
- 1 toilet paper tube (per group)
- Graph paper
- Tape

Student Pages

- Seasonal Solution (lab sheet)
- "The 20-Year Winter" (article)
- Earth in Space (thinking visually)

Exploration/Pre-Reading

In this exploration, students will be given supplies to create a model of Earth's movement around the Sun so they can look for a way to generate seasons. Before class, you will need to prepare the model materials for each group.

> Sun: Several options will work here. You can use a lamp with the shade removed. You can remove the shade from a nightlight and plug it into an extension cord. You can also make a small circuit from a battery pack (or batteries taped end-to-end) with lightbulbs cut from an old string of holiday lights. (see Figure 14.1). In any case, you will need to set the Sun on a stand so that students can move the Earth model around it. I used a toilet paper tube for this purpose (Figure 14.2, p. 154).

Figure 14.1. Two Possible Lightbulb Configurations

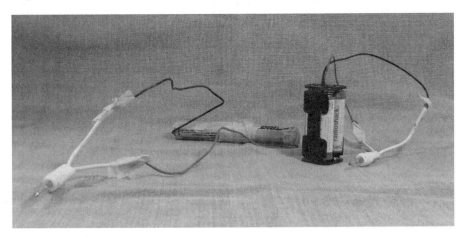

Figure 14.2. Stand for the Sun

Earth: Place a 3 in. Styrofoam ball on a thin dowel. The dowel should be inserted so that the Earth is oriented straight up and down, not reflecting the tilt of the Earth. Also, the Earth should be positioned so that it is in the same plane as the Sun if the end of the stick is touching the table (Figure 14.2). Use a permanent marker to draw an equator around the center of the ball, and roughly mark the location of Washington, DC, and Santiago, Chile, (roughly 35° above and below the equator). Likewise, color a small circle on the top to represent the North Pole.

For the light intensity experiment (the "hint"), tape a ruler or dowel so that it extends about 15 cm from the end of each flashlight and provide graph paper.

Begin the activity by asking students about possible causes for the seasons, and record those options in a list. Initially, don't affirm the right answer; simply list all possibilities that students think up.

Then introduce the data showing the opposite seasons in Santiago and Washington, DC. Explain that the students' challenge is to create a model that can explain the seasonal changes. Help them get started using the model materials. Explain that the dowel coming from Earth must touch the table at all times (to keep Earth in the same plane as the Sun). Ask them to use the models to show day and night. Then ask them to use their models to show how Earth orbits the Sun. This should be enough of an introduction for them to start experimenting with how to move Earth so that light hits at different places at different times.

After a few minutes of experimenting, have students work through the hint provided. Even if students arrive quickly at the correct answer, have them complete the hint section for a strong visual of the difference between direct and indirect sunlight.

For the challenge questions, you will need to point out that the model they are using is only a partial representation (it does not incorporate seasonal lag). In fact, the dark and light months in the Arctic occur about a month earlier (in December and June) than this model predicts.

Introduce the Reading. Tell students that as they read, they should compare what this article says about how seasons occur to what they saw in their models.

Reading Strategy: Previewing Diagrams and Illustrations

Note that this strategy was introduced in Chapter 13 (p. 133). If you have not used Chapter 13 with your class, tell students that in some books that they read, the pictures are extras. In science writing, however, the pictures and diagrams often carry important information. Looking at pictures and making predictions about what they mean before reading can make the text easier to understand.

If you have already introduced this strategy, remind students about the importance of pictures in science texts, and tell them they are going to practice three steps that they can use to preview diagrams and illustrations:

- Step 1: Describe what you see in the diagram, without worrying about whether you know the correct terms for everything shown.
- Step 2: Make a prediction about what the diagram or illustration shows.
- Step 3: Come up with at least one question about the diagram that might be answered in the text.

Place students into their reading groups, and have them preview Figure S14.1 (p. 159), with the leader doing step 1, the emergency manager doing step 2, and the interpreter doing step 3. Pause and ask the class to share some of their thoughts and observations. Do not confirm or contradict their predictions—tell them they will have to read the text to find out if they are correct. Repeat for Figures S14.2 and S14.3 (p. 160), with each student doing a different step each time.

Journal Question

When you looked at the diagrams before reading, your group discussed three questions: (1) What do you see in the diagram? (2) What do you predict the diagram is illustrating? (3) What question do you think the text will answer about the diagram? Which of these three questions was most helpful for understanding the text? Why?

FIND OUT MORE
To learn more about seasons across the solar system, see the solar system exploration site by NASA at *http://solarsystem.nasa. gov.*

Application/Post-Reading

- Thinking Visually: Earth in Space
- Writing Prompt: A classmate incorrectly claims that winter is when Earth is farthest away from the Sun in its orbit, and summer happens when Earth is closest to the Sun. Make a correct claim about what causes seasons and support it with evidence and reasons from your lab and the article. In your response, provide evidence that rebuts your classmate's claim.
  - Prewriting Questions: What is a rebuttal? How can you include a rebuttal when writing about a claim? What kinds of writing words might you use in a rebuttal? (*If ... then* statements, *however*, and *therefore* are useful words for rebuttals.)
  - Key Evaluation Point: Seasons are caused by the tilt of Earth, which leads to shorter daylight hours and less direct sunlight during winter. It cannot be caused by distance from the Sun because Earth's orbit is nearly circular, and because that model would mean that all places on Earth should have seasons at the same time.

Seasonal Solution

Santiago, Chile, and Washington, DC, are located near the same longitude line. However, Santiago is about 35° below the equator, and Washington, DC, is about 35° above the equator. As you can see in Table S14.1, the timing of their seasons is very different.

Table S14.1. Seasons in Santiago, Chile, and Washington, DC

Location	January			July		
	Average High	Average Low	Season	Average High	Average Low	Season
Santiago, Chile	86	62	Summer	42	29	Winter
Washington, DC	57	37	Winter	88	72	Summer

Scientists often try out different models to see which one might explain data they have gathered. You goal is to create a model of the Earth and Sun that can explain the difference in the timing of seasons in Santiago and Washington, DC.

Your model also has to match other information that we have about the movement of Earth, including that it orbits the Sun, that it rotates, and that it stays in the same plane as the Sun.

1. When you have arrived at a solution, draw your arrangement for January and July (on a separate piece of paper).

2. Describe how your model accounts for the seasonal differences between Santiago and Washington, DC.

   Hint: Explore light intensity for a few minutes.

   A. Point your flashlight straight down at the graph paper provided, so that the stick touches the page. Draw a circle around the brightest part of the light. How many squares were lit?_____

   B. Keeping the stick touching the paper, tilt the flashlight about halfway to the table. Draw around the brightest part of the light. How many squares were lit?_____

   C. The same amount of energy was hitting the bright area both times. Think about just one square. Would it get more energy when the light hit it straight on, or when the light came at an angle? _____

   D. How could you use this idea in your Earth and Sun model?_____

   _____

   _____

Challenge

The Arctic Circle has a period of nighttime that lasts for a little over a month each year. During that time, the Sun never rises. At another time, the Arctic Circle has a period of daylight that also lasts over a month. Use your model to predict approximately when those periods of darkness and light should take place. (The Antarctic Circle should have darkness and light at opposite times.)

Estimated dark month for Arctic Circle: _____

Estimated light month for Arctic Circle: _____

(*Note:* The model you are using is only a partial representation of the seasons. In fact, the dark and light months in the Arctic occur about a month earlier than your model predicts.)

The 20-Year Winter

Imagine if winter lasted for 20 years. If it started when you were in middle school, you would be in your 30s before spring rolled around. On Uranus, not only does winter last for 20 Earth years, but the temperature can drop to −300°F. When spring finally comes and the atmosphere begins to warm, the storms are a sight to behold, even from across the solar system. Astronomers have viewed spring storms that are as large as the entire continent of North America.

Seasons

Seasons on Uranus are the result of the same factors that cause seasons on Earth. We think of the Earth as being straight up and down, with the North Pole on top and the South Pole on the bottom. However, the Earth is tilted 23.4° relative to the Sun. This means that as it orbits, the part of the Earth pointing toward the Sun changes. When one of the poles is angled toward the Sun, it gets more sunlight and experiences summer. When that same pole is angled away from the Sun, it gets less sunlight and experiences winter. The middle part of the Earth, near the equator, gets about the same amount of sunlight all year, and the seasons are not as noticeable.

Although Earth's orbit is technically an ellipse, it is just slightly elliptical. It is actually very close to being a circle. Therefore, the Earth as a whole stays about the same distance from the Sun all year long. It is the tilt that causes changes in sunlight over the year.

Solstices and Equinoxes

If you live in North America, you have probably noticed that winter is not all about being cold. The Sun comes up later in the morning and sets sooner at night, which gives fewer hours of daylight. During those daylight hours, the sunlight isn't as bright. The Sun's rays come in at an angle, so the energy from the Sun is spread out over a greater area (Figure S14.1).

In the Northern Hemisphere, the darkest day of the year comes around December 22, when the North Pole points away from the Sun. This day is called winter solstice in the Northern Hemisphere. On that same day, the South Pole points toward the Sun, and this

Figure S14.1. The Intensity of the Sun's Rays at Different Times of Year

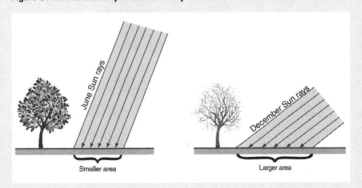

is called summer solstice in the Southern Hemisphere. On about June 22, the poles are reversed, and the Northern Hemisphere experiences summer solstice while the Southern Hemisphere has winter solstice. Figure S14.2 shows how the Sun's rays hit the Earth over the course of a year to produce the solstices.

In between the solstices, there are two days in which the two hemispheres get equal amounts of Sun. These days are called equinoxes. During an equinox, neither pole points toward the Sun.

Figure S14.2. The Seasons

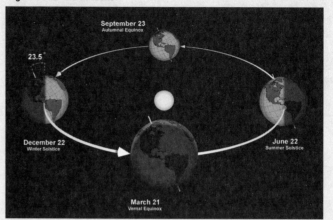

Seasonal Lag

The coldest days of the year usually come a little after the winter solstice, even though the amount of sunlight is increasing. For example, the coldest day of winter in Washington, DC, is usually around January 19. This delay is called seasonal lag. Seasonal lag occurs because both water and land absorb heat during the warm months. As the temperatures cool, the water and land gradually release their heat into the atmosphere. When the water and land finally cool off, it takes time for them to heat back up again. For this reason, the hottest days in the Northern Hemisphere are usually in July, even though the summer solstice occurs in June.

So What Does This Have to Do With Uranus?

Uranus also orbits the Sun in an orbit that is almost a circle. It is also tilted on its axis. However, Uranus is tilted 98° relative to the Sun (Figure S14.3). It is almost lying on its side. On Uranus, whichever pole points toward the Sun gets almost all of the sunlight. Whichever pole points away from the Sun is in total darkness. When you combine that tilt with Uranus' long journey around the Sun—equal to 84 Earth years—you have a recipe for a nasty winter season.

The long seasons mean that modern astronomers are just getting their first glimpses of spring on the Northern Hemisphere of Uranus. The last time the Northern Hemisphere had spring was in 1923. Telescopes have come a long way since 1923, and astronomers are excited about what they might learn.

Figure S14.3. The Axial Tilts of Earth and Uranus

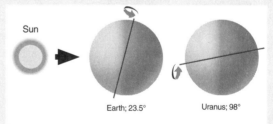

THE BIG QUESTION

What factors cause seasons on both Earth and Uranus?

Earth in Space

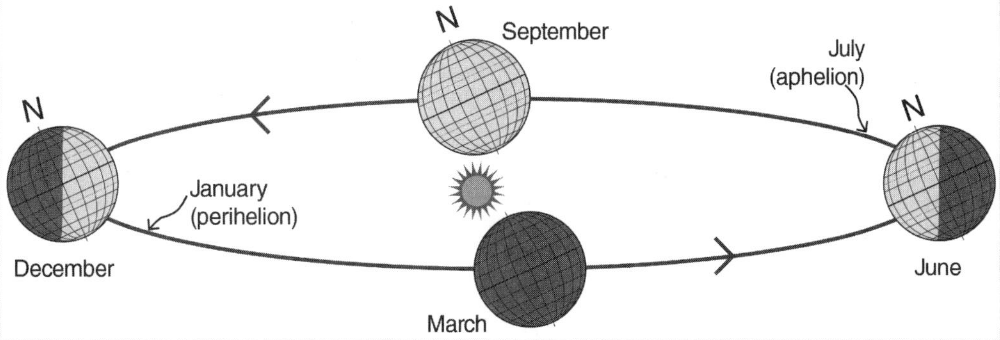

1. Write a caption for this diagram that explains what it is illustrating.

2. Are the parts of the diagram shown to scale? How do you know?

3. If you were looking down on the North Pole, would the Earth move clockwise or counterclockwise?

4. The perihelion is the point in Earth's orbit where it is *closest* to the Sun. What season is it in North America at the perihelion?

5. Based on the diagram, what do you think the word "aphelion" means?

6. Use the information from the article to explain in your own words what a *solstice* is.
   - Label the places in the diagram that show solstices.

7. Use the information from the article to explain the meaning of *equinox.*
   - Label the places in the diagram that show equinoxes.

Hair Dryer Helper

Chapter 15

Topics
- Sources of energy
- Energy policy
- Effects of population on resource use

Reading Strategy
- Evaluating persuasive science writing

Lesson Objectives: Connecting to National Standards

The following list shows the *Next Generation Science Standards (NGSS)* and *Common Core State Standards (CCSS)* supported by this activity.

NGSS: *Science and Engineering Practices*
- Planning and Carrying Out Investigations
- Obtaining, Evaluating, and Communicating Information

NGSS: *Disciplinary Core Ideas*
- **ESS3.A.** Natural Resources
- **ESS3.C.** Human Impacts on Earth Systems

NGSS: *Crosscutting Concept*
- Patterns

CCSS: *Literacy in Science and Technical Subjects*
- **CCSS.ELA-Literacy.RST.6-8.2.** Determine the central ideas or conclusions of a text; provide an accurate summary of the text distinct from prior knowledge or opinions.
- **CCSS.ELA-Literacy.RST.6-8.5.** Analyze the structure an author uses to organize a text, including how the major sections contribute to the whole and to an understanding of the topic.
- **CCSS.ELA-Literacy.RST.6-8.6.** Analyze the author's purpose in providing an explanation, describing a procedure, or discussing an experiment in a text.
- **CCSS.ELA-Literacy.WHST.6-8.9.** Draw evidence from informational texts to support analysis, reflection, and research.

Background

In this chapter, students are asked to consider public policy decisions as they relate to energy usage. They will also consider how to write and evaluate persuasive pieces. These activities are best suited for the end of a unit on renewable and nonrenewable energy sources, because the article assumes a basic understanding of energy sources and climate change.

Materials

- Electricity usage meter (such as Kill-A-Watt)
- 4–5 hair dryers
- "Wigs" made of five to six short strands of yarn taped together
- Sanitized safety glasses or goggles

Student Pages

- Hair Dryer Helper (lab sheet)
- "Mandate Energy Efficiency!" (article)
- Analyzing the Argument (to accompany the article)
- Energy Use Around the World (thinking visually)

Exploration/Pre-Reading

For this activity, students will serve as consumer advocates to evaluate and rate hair dryers. In the process, they will observe that hair dryers can vary greatly in how much energy they use, and that this might be a feature that consumers should know about their appliances. Hair dryers work well for this activity because they are small and portable, yet they use large amounts of energy quickly. In just three to four minutes, the difference in usage is visible on an electricity usage meter.

To conduct the energy analysis, students will need access to an electricity usage meter, such as the Kill-A-Watt meter. These can be purchased for around $25 from a hardware store or online. You can organize the activity to make do with just one meter, or you can provide one to each group. The class will also need three to four different hair dryers. You will see the most difference in energy use if there are representatives from at least two of these categories: full size, compact/travel, and "ecofriendly."

To begin, students will brainstorm a list of features they might like to include in their report. You may want to show students a sample consumer evaluation done by a group such as Consumer Reports. They will be able to see how a purchase can be rated across a variety of features.

Students may want to consider price (if you know it); options such as high, low, and cooling; how fast the hair dryer dries a wig; how easy it is to plug in; how comfortable it is to use; or available attachments such as diffusers. Students may not think to measure the hair dryer's energy usage, so you may need to introduce this aspect.

Then students will decide on specific tests or rating systems for three to five of those features (with one being energy use). They should agree on the exact procedure to be used to rate each hair dryer. Depending on what tests they design, you might have each group do all tests for one hair dryer, or you might have each group be responsible for one of the evaluations and have them pass the hair dryers from group to group.

The class should then create a data chart to show all their findings. Finally, individuals or groups can write a report explaining which hair dryer they recommend, and why.

Introduce the Reading. Tell students they are going to read an argument in which an author is going to try to convince them of a particular point of view. This is a type of persuasive writing, as they might have discussed in language arts classes. As you read an argument, you should evaluate whether you agree with the author and why.

SAFETY NOTE
Students doing this sort of lab should wear eye protection (sanitized safety glasses or goggles meeting the ANSI Z87.1 standard) during the setup, hands-on investigation, and takedown. Use caution when working around hair dryers. They are hot and can burn the skin. Keep water away from any electrical equipment to prevent shock. Use only GFI-protected electrical circuits for hair dryers.

TEACHING TIP
If you are using hair dryers that were purchased recently, you might be able to get your school newspaper to print some of the hair dryer evaluations.

Reading Strategy: Evaluate Arguments, Generate a Diagram of the Flow of the Argument

Begin by reviewing what students may know about argumentation (discussed in more detail in Chapter 3, p. 25). In the most basic argument, the author makes a claim, supports it with reasons, and says why those reasons support the claim. In a more complex argument, the author may also offer counterclaims, or reasons other people might disagree. The author will then rebut the counterclaims and say why those people are incorrect. It can be helpful to show students a diagram of this type of argument, as shown in Figure 15.1. The argument that students are going to read will have two counterclaims.

Figure 15.1. Diagram of an Argument

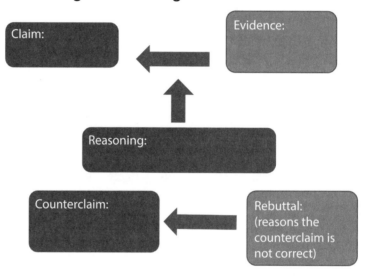

Explain to students that there are three basic steps to evaluating an argument.

1. Make sure you understand the argument.
2. Question the argument.
3. Decide whether to accept all, part, or none of the argument.

They will complete the first step after they read the article; you will work on the other two steps as a class. Have students read the article and complete the worksheet "Analyzing the Argument."

When they finish, engage the class in a discussion around the following questions (step 2):

- Do the "facts" presented seem to be accurate? (Discuss the sources provided with your students; you may need to show them how to use the superscript numbers to find the footnotes. What kind of sources are legitimate? How could they find out if the information given really came from those sources?)
- Can you think of any counterclaims the author might have overlooked? Are they valid? (You might point out that the author only gives examples of instances in which government regulation was effective. A quick search on lightbulb regulation will show an instance in which the newly regulated lightbulbs [CFC's] turned out to generate toxic waste and not last as many years as initial reports had promised.)
- Is there anything else you need to find out before accepting the argument? If so, what? (Point out that it is often wise to find something written by someone with another point of view, to see how they argue their claim.)

The last step is to decide what parts of the argument to accept. Point out that students can agree completely with the author or accept only some of her argument. In this case, for example, students might agree with her statement of the problem but disagree with her solution. Alternatively, they might agree that her solution is good in some instances but not in others. The writing prompt for this chapter asks students to evaluate the claim made by the author of this article.

Journal Question

Why do you think it is important to evaluate an author's argument when you read persuasive writing?

Application/Post-Reading

- Thinking Visually: Energy Use Around the World
- Writing Prompt: What parts of the author's argument in "Mandate Energy Efficiency!" do you accept? Did the author support

> **FIND OUT MORE**
> To learn more about energy efficiency standards in the United States, see the website of the U.S. Office of Energy Efficiency at *http://energy.gov/eere/buildings/appliance-and-equipment-standards-program.*

her position well? Why or why not? What other ideas should she have considered?

- o Prewriting Activity: Use the class discussion to help students think about what they might write.
- o Key Evaluation Point: Students should consider whether data have been provided to back up the claim as well as the source of that data. They should address whether the counterclaims have been rebutted convincingly.

Hair Dryer Helper

1. You have been hired as a consumer advocate to rate hair dryers. Your goal is to evaluate the dryers so that people will know which hair dryer they should buy. To begin, brainstorm a list of things that you would look for in a good hair dryer.

2. Narrow your ideas to three to five features on which you will rate the hair dryers. Decide on a specific way to measure how each hair dryer scores on each feature. Some features may lead to a number score. Others may just be rated as "yes" or "no." Be sure to consider the exact conditions for each test. For example, if you are drying a wig, how will you wet it and with how much water?

3. Create a data chart to show the score for each hair dryer on each feature.

4. Working with your classmates, test each hair dryer on each feature and complete the data chart.

5. Make a claim: Which hair dryer would you recommend to consumers? Explain what evidence you used to decide and why you selected that evidence. Write as if you are explaining your choice in a magazine article.

Hair Dryer Helper

Mandate Energy Efficiency!

We have an energy problem. Ninety percent of the energy used in the United States comes from oil, natural gas, coal, or nuclear power.[1] All of these energy sources are nonrenewable. This means that once Earth's supply of these fuels is used up, they are gone forever. These fuels have other disadvantages as well. When oil and coal are burned, they release toxic air pollution. Oil, coal, and natural gas all contribute greenhouse gases to the atmosphere and help drive climate change. Nuclear reactors generate radioactive waste that remains dangerous for thousands of years.

Governments and researchers around the world have been working to increase the amount of energy produced from renewable resources, such as windmills, solar power, and hydroelectric plants. However, these energy sources tend to be more expensive than nonrenewable sources, and they can also have environmental drawbacks. For example, manufacturing solar panels can produce toxic waste. Likewise, hydroelectric plants can interfere with the ecosystems in the rivers where they operate (Figure S15.1).

Figure S15.1. Hydroelectric Power Plant Dam

Clearly, reducing energy use must be a part of any future energy solution for the world. Educational programs can teach people to turn off electronics and appliances when they are not needed and to drive less or carpool. However, there is a limit to how much an individual can reduce his or her energy use and still live in the modern world. The most effective way to reduce the demand for energy is to force manufacturers to make appliances and cars that require less of it.

For many products, more energy efficient options are available—but at a cost. Factories have to be redesigned to build the new products. Sometimes, more research must be done. Suppose a company produces hair dryers (Figure S15.2). This company might not want to spend the money to make its hair dryers more energy efficient because it fears that other companies will sell cheaper hair dryers that are not energy efficient. Most people would buy the cheaper hair dryers. Governments can help by requiring all hair dryer companies to make hair dryers that meet the same energy efficiency standards. While hair dryer prices may go up in the short term for all companies, once the manufacturing changes are over, the price often goes back down.

Figure S15.2. A Hair Dryer

170

Table S15.1. Comparison of Energy Consumption of Old and New Refrigerators

Year Refrigerator Was Made	Average Electricity Used in a Year	Annual Cost of Electricity (based on the average cost of electricity in the United States for 2015: 0.111 per kWh)
Before 1980	2,215 kWh	$245.87
1980–1989	1,709 kWh	$189.70
1990–1992	1,285 kWh	$142.64
1993–2000	857 kWh	$95.13
2001–2008	537 kWh	$59.61

Source: Data from Refrigerator Retirement Savings Calculator at EnergyStar.gov.
Note: kWh = kilowatt hour.

It has been argued that energy efficiency standards prevent consumers from being able to buy what they want. In this argument, energy efficient products cost more, and consumers should have the right to pay less if they don't care about energy efficiency. This argument is wrong on three counts. First, energy efficient products are not always more expensive (Table S15.1). For example, energy efficiency standards for refrigerators were first introduced in 1973. Today's refrigerators use only a third of the energy used in 1973 and cost about half what older fridges cost.[2] Second, pollution and climate change affect everyone. People who prefer cheaper appliances shouldn't be able to force those effects on everyone else. Third, it costs all of us to continue building new power plants to keep up with the need for more electricity. We all save money if we do not have to pay taxes or higher prices on energy to pay for new power plants.

Critics also claim that when people save energy in one area, they simply go use more energy elsewhere, in a process called the rebound effect. For example, if a car uses less gas, driving becomes less expensive, so people do more driving. However, this rebound effect is not as big as critics suggest. A recent article, published in the journal Nature[3], analyzed a number of studies of the rebound effect. The researchers found that people used only 5–30% of their energy savings elsewhere. Therefore, a driver who saves 100 gallons of gas in a fuel-efficient car wastes at most 30 gallons on extra driving. At least 70 gallons of gas are still saved.

For these reasons, governments should set energy efficiency standards for cars and appliances, and increase those standards over time so that companies continue working toward peak energy efficiency.

Endnotes

1. U.S. Energy Information Administration. 2014. *Monthly energy review—February.* Table 1.3, preliminary 2013 data. *www.eia.gov/totalenergy/data/monthly/pdf/mer.pdf.*
2. U.S. Office of Energy Efficiency and Renewable Energy. 2014. History and impact of appliance and equipment standards. *www.energy.gov/eere/buildings/history-and-affects.*
3. Gillingham, K., M. J. Kotchen, D. S. Rapson, and G. Wagner. Energy policy: The rebound effect is overplayed. *Nature* 493.7433 (2013): 475–76.

THE BIG QUESTION

What does the author of this article want people to do?

Analyzing the Argument

The author's position: the author begins by describing a problem and a solution.

    1. What problem is the author concerned about?

    2. What solution does the author want you to support?

    3. What reason does the author give to explain why this is a good solution?

Counterclaims: The author describes two arguments that people might make against her solution.

    4. What is the first counterclaim (paragraph five)?

    5. What reason does the author give for rejecting this claim (her rebuttal)?

    6. What is the second counterclaim (paragraph six)?

    7. What reason does the author give for rejecting this claim (her rebuttal)?

Energy Use Around the World

Country	Total Population (rounded to nearest thousand)	Energy Use per Person (in million Btu)	National Energy Use (in million Btu)
Sweden	9,074,000	244.879	2,222,000,000
Kenya	40,843,000	5.383	22,000,000
United Kingdom	62,348,000	142.971	8,914,000,000
Japan	127,579,000	170.660	21,773,000,000
United States	309,330,000	316.946	98,041,000,000
India	1,173,108,000	18.685	21,920,000,000
China	1,330,141,000	75.843	100,881,000,000

Source: Data from the U.S. Energy Information Administration for the year 2010. *www.eia.gov/cfapps/ipdbproject.*
Note: Btu = British thermal unit.

1. Which has more people, the United Kingdom or Sweden? Which one uses more energy?

2. Compare the data given for Japan and India. What do you notice? Use the phrase "even though" in your response.

3. Compare the data given for the United States and China. Describe at least one way their data are similar and one way their data are different.

4. Select one other pair of countries from the chart to compare. Explain what makes that comparison interesting to you.

5. Do countries with larger populations always use more energy? Use examples to support your answer.

6. Imagine that Kenya has a big year for babies and 1 million more children are born than usual! In the same year, the United States *also* has 1 million extra babies. Which children will have a greater effect on global energy use?

7. We often think of population growth as a problem for the developing world. Why is it also a problem that the United States should consider?

Connections to the Next Generation Science Standards and the Common Core State Standards

Chapter 4

Standard
MS-ESS1. Earth's Place in the Universe (*www.nextgenscience.org/msess1-earth-place-universe*)

Performance expectations
The materials/lessons/activities outlined in this chapter are just one step toward reaching the performance expectations listed below.

MS-ESS1-4. Construct a scientific explanation based on evidence from rock strata for how the geologic timescale is used to organize Earth's 4.6-billion-year-old history.

Dimension	Element	Matching Student Task or Question From the Activity
Science and engineering practices	• Planning and Carrying Out Investigations • Analyzing and Interpreting Data • Engaging in Argument From Evidence	• Students design and carry out an experiment to see how speed affects distance between footprints. • Students use data from their own experiment, along with data from a variety of animals, to support their claim. • Students make a claim about how stride length is affected by speed and support it with evidence. • Students read about a claim made by a paleontologist and analyze the evidence he provides for that claim. • In the thinking visually section, students identify the assumptions made by the researchers in using modern animals as models for dinosaurs.
Disciplinary core ideas	**ESS1.C.** The History of Planet Earth • The geologic timescale interpreted from rock strata provides a way to organize Earth's history. Analyses of rock strata and the fossil record provide only relative dates, not an absolute scale.	• Although this chapter deals indirectly with this core idea, it provides an important foundation for learning about geologic timescales, namely that "the assumption that theories and laws that describe the natural world operate today as they did in the past and will continue to do so in the future"(NGSS Lead States, p. 68; constructing an explanation).
Crosscutting concepts	• Patterns • Stability and Change	• Students look for patterns in speed versus stride length in their own data and data from a variety of other animals. • Students construct an argument and then analyze an argument made by practicing scientists based on the idea that some processes remain stable across geologic time.
CCSS Correlations		
Reading standard(s)	• **CCSS.ELA-Literacy.RST.6-8.1.** Cite specific textual evidence to support analysis of science and technical texts. • **CCSS.ELA-Literacy.RST.6-8.2** Determine the central ideas or conclusions of a text; provide an accurate summary of the text distinct from prior knowledge or opinions. • **CCSS.ELA-Literacy.RST.6-8.5.** Analyze the structure an author uses to organize a text, including how the major sections contribute to the whole and to an understanding of the topic. • **CCSS.ELA-Literacy.RST.6-8.6.** Analyze the author's purpose in providing an explanation, describing a procedure, or discussing an experiment in a text.	• *Reading strategies:* Comprehension coding and reading in groups (both of which require citing specific chunks of text) • Comprehension coding requires students to note when their prior knowledge conflicts with what the text says. • Students outline the argument made by a practicing scientist about the purpose of a dinosaur's skull structure and analyze its effectiveness.
Writing standard(s)	• **CCSS.ELA-Literacy.WHST.6-8.1.** Write arguments focused on discipline-specific content.	• Students write an argument with a claim and evidence for their interpretation of a dinosaur trackway.

Chapter 5

Standard		
MS-ESS2. Earth's Systems (*www.nextgenscience.org/msess2-earth-systems*)		
Performance expectations		
The materials/lessons/activities outlined in this chapter are just one step toward reaching the performance expectations listed below.		
MS-ESS2-2. Construct an explanation based on evidence for how geoscience processes have changed Earth's surface at varying time and spatial scales.		

Dimension	Element	Matching Student Task or Question From the Activity
Science and engineering practices	• Developing and Using Models • Constructing Explanations and Designing Solutions	• Students use a model of a mountain to examine causes and effects of erosion. • Students use provided materials to explore possible solutions to an erosion problem.
Disciplinary core ideas	**ESS2.C.** The Roles of Water in Earth's Surface Processes • Water's movements—both on the land and underground—cause weathering and erosion, which change the land's surface features and create underground formations.	• Students observe and read about the ways water and gravity drive erosion and deposition activity on Earth.
Crosscutting concepts	• Cause and Effect • Stability and Change	• Students experiment with causes and effects of erosion using a model. • Students observe and read about how erosion changes the Earth, and how humans attempt to create stability.
CCSS Correlations		
Reading standard(s)	• **CCSS.ELA-Literacy.RST.6-8.4.** Determine the meaning of symbols, key terms, and other domain-specific words and phrases as they are used in a specific scientific or technical context relevant to grades 6–8 texts and topics. • **CCSS.ELA-Literacy.RST.6-8.9.** Compare and contrast the information gained from experiments, simulations, video, or multimedia sources with that gained from reading a text on the same topic.	• *Reading strategy:* Finding the meaning of new words • The Big Question in this chapter requires students to compare their lab results to their reading and use new words from the reading to describe what happened in lab.
Writing standard(s)	• **CCSS.ELA-Literacy.WHST.6-8.2.** Write informative/explanatory texts, including the narration of historical events, scientific procedures/experiments, or technical processes.	• *Writing prompt:* Digger Johnson has just gotten a contract to build a road into the side of a mountain. He doesn't know a thing about erosion. Write Mr. Johnson a letter explaining how erosion could affect his road. Give him some suggestions for protecting it.

Once Upon an Earth Science Book

Chapter 6

Standard		
MS-ESS2. Earth's Systems (*www.nextgenscience.org/msess2-earth-systems*)		

Performance expectations

The materials/lessons/activities outlined in this chapter are just one step toward reaching the performance expectations listed below.

MS-ESS2-2. Construct an explanation based on evidence for how geoscience processes have changed the Earth's surface at varying time and spatial scales.

MS-ESS2-3. Analyze and interpret data on the distribution of fossils and rocks, continental shapes, and seafloor structures to provide evidence of the past plate motions.

Dimension	Element	Matching Student Task or Question From the Activity
Science and engineering practices	• Analyzing and Interpreting Data • Engaging in Argument From Evidence	• At three stations, students analyze evidence from mountain ranges, glaciation, and fossils that Wegener used to propose his theory of moving continents. • Students make and support a claim as to whether torn paper originated from the same advertisement. • Students read about how scientists received Wegener's claims about continental drift. • *Writing prompt.* Imagine that you could go back in time and talk to the geologists at the conference where they mocked Wegener's idea. Explain to them how new evidence from the ocean and satellites supports the idea that continents can move.
Disciplinary core ideas	**ESS1.C.** The History of Planet Earth • Tectonic processes continually generate new ocean sea floor at ridges and destroy old sea floor at trenches. **ESS2.B.** Plate Tectonics and Large-Scale System Interactions • Maps of ancient land and water patterns, based on investigations of rocks and fossils, make clear how Earth's plates have moved great distances, collided, and spread apart.	• In the thinking visually section, students analyze a diagram of sea floor spreading. • At three stations, students examine evidence from mountain ranges, glaciation, and fossils that Wegener used to propose his theory of moving continents. They read an article about Wegener's claims and how our understanding of plate tectonics has changed over time.
Crosscutting concepts	• Cause and Effect	• Students look at data on the effects of plate tectonics, and then read about the process scientists used to determine cause.
CCSS Correlations		
Reading standard(s)	• **CCSS.ELA-Literacy.RST.6-8.2.** Determine the central ideas or conclusions of a text; provide an accurate summary of the text distinct from prior knowledge or opinions. • **CCSS.ELA-Literacy.RST.6-8.7.** Integrate quantitative or technical information expressed in words in a text with a version of that information expressed visually (e.g., in a flowchart, diagram, model, graph, or table).	• *Reading strategy.* Chunking • Chunking requires focusing on the text in small bits to ensure that each piece is understood. • Students study diagrams at stations before reading; additional diagrams are provided in the text.
Writing standard(s)	• **CCSS.ELA-Literacy.WHST.6-8.1.** Write arguments focused on discipline-specific content. • **CCSS.ELA-Literacy.WHST.6-8.2.** Write informative/explanatory texts, including the narration of historical events, scientific procedures/experiments, or technical processes.	• *Writing prompt.* Imagine that you could go back in time and talk to the geologists at the conference where they mocked Wegener's idea. Explain to them how new evidence from the ocean and satellites supports the idea that continents can move. • *The Big Question:* Do scientists ever change their minds about how something on Earth works? What helped scientists eventually accept Wegener's claim?

Chapter 7

Standard		
MS-ESS2. Earth's Systems (*www.nextgenscience.org/msess2-earth-systems*)		

Performance expectations

The materials/lessons/activities outlined in this chapter are just one step toward reaching the performance expectations listed below.

MS-ESS2-2. Construct an explanation based on evidence for how geoscience processes have changed Earth's surface at varying time and spatial scales.

Dimension	Element	Matching Student Task or Question From the Activity
Science and engineering practices	• Developing and Using Models • Constructing Explanations and Designing Solutions	• Students model the formation of a landform in which sea life fossils end up on the top of a mountain. • Using information from their model, students respond to the following: Imagine you are hiking this mountain with your friend and see one of the colored shells. She asks, "How did a fossil shell get on top of a mountain?" What do you tell her?
Disciplinary core ideas	**ESS1.C.** The History of Planet Earth • The geologic timescale interpreted from rock strata provides a way to organize Earth's history. Analyses of rock strata and the fossil record provide only relative dates, not an absolute scale.	• Students read an article on how scientists date landforms (with specifics from the Burgess Shale). They use the information to respond to the following prompt: Write a paragraph that explains the difference between an absolute age and a relative age. Give an example of each from geology, and then give an example of how you could describe how long you've been alive using a relative age and an absolute age.
Crosscutting concepts	• Systems and System Models • Patterns	• Students use a model to explore changes to a landform over a large area and through a long period of time. • In thinking visually, students interpret patterns in rock formations and see how those patterns can indicate the history of those rocks.
CCSS Correlations		
Reading standard(s)	• **CCSS.ELA-Literacy.RST.6-8.4.** Determine the meaning of symbols, key terms, and other domain-specific words and phrases as they are used in a specific scientific or technical context relevant to grades 6–8 texts and topics.	• *Reading Strategy:* Finding the meaning of new words
Writing standard(s)	• **CCSS.ELA-Literacy.WHST.6-8.2.** Write informative/explanatory texts, including the narration of historical events, scientific procedures/experiments, or technical processes. • **CCSS.ELA-Literacy.WHST.6-8.2.c.** Use appropriate and varied transitions to create cohesion and clarify the relationships among ideas and concepts.	• Write a paragraph that explains the difference between an absolute age and a relative age. Give an example of each from geology, and then give an example of how you could describe how long you've been alive using a relative age and an absolute age. • *Prewriting questions:* What science words will you need? What writing words could you use? (comparison and contrast words such as *similarly, in contrast, however,* etc.).

Chapter 8

Standard
MS-ESS2. Earth's Systems (*www.nextgenscience.org/msess2-earth-systems*)

Performance expectations
The materials/lessons/activities outlined in this chapter are just one step toward reaching the performance expectations listed below.

MS-ESS2-1. Develop a model to describe the cycling of Earth's materials and the flow of energy that drives this process.

Dimension	Element	Matching Student Task or Question From the Activity
Science and engineering practices	• Engaging in Argument From Evidence	• Students make an argument for which location police should search for the stolen goods, based on evidence from rock samples.
Disciplinary core ideas	**ESS2.A.** Earth's Materials and Systems • All Earth processes are the result of energy flowing and matter cycling within and among the planet's systems. This energy is derived from the Sun and Earth's hot interior. The energy that flows and matter that cycles produce chemical and physical changes in Earth's materials and living organisms.	• Students observe several types of rocks and read about rock types and how they can influence a police investigation.
Crosscutting Concepts	• Energy and Matter: Flows, Cycles, and Conservation	• In the thinking visually section, students answer questions about the cycling of matter through rocks.
CCSS Correlations		
Reading standard(s)	• **CCSS.ELA-Literacy.RST.6-8.1.** Cite specific textual evidence to support analysis of science and technical texts. • **CCSS.ELA-Literacy.RST.6-8.9.** Compare and contrast the information gained from experiments, simulations, video, or multimedia sources with that gained from reading a text on the same topic.	• *Reading strategy:* Chunking • Chunking involves close reading of dense text in order to understand each idea presented.
Writing standard(s)	• **CCSS.ELA-Literacy.WHST.6-8.1.** Write arguments focused on discipline-specific content. • **CCSS.ELA-Literacy.WHST.6-8.1.a.** Introduce claim(s) about a topic or issue, acknowledge and distinguish the claim(s) from alternate or opposing claims, and organize the reasons and evidence logically.	• Students write an argument for which location police should search for the stolen goods, based on evidence from rock samples. • *Extension:* If your students are accustomed to writing about claims and evidence, help them include counterclaims in their response with this demonstration.

Chapter 9

Standards
MS-ESS2. Earth's Systems (*www.nextgenscience.org/msess2-earth-systems*)
MS-ESS3. Earth and Human Activity (*www.nextgenscience.org/msess3-earth-human-activity*)

Performance expectations

The materials/lessons/activities outlined in this chapter are just one step toward reaching the performance expectations listed below.

MS-ESS2-4. Develop a model to describe the cycling of water through Earth's systems driven by energy from the Sun and the force of gravity.

MS-ESS3-1. Construct a scientific explanation based on evidence for how the uneven distributions of Earth's mineral, energy, and groundwater resources are the result of past and current geoscience processes.

Dimension	Element	Matching Student Task or Question From the Activity
Science and engineering practices	• Developing and Using Models	• Students use a model to look at the effect of water use on groundwater storage.
Disciplinary core ideas	**ESS2.A.** Earth's Materials and Systems • All Earth processes are the result of energy flowing and matter cycling within and among the planet's systems. This energy is derived from the Sun and Earth's hot interior. The energy that flows and matter that cycles produce chemical and physical changes in Earth's materials and living organisms. **ESS2.C.** The Roles of Water in Earth's Surface Processes • Water continually cycles among land, ocean, and atmosphere via transpiration, evaporation, condensation and crystallization, and precipitation, as well as downhill flows on land. • Water's movements—both on the land and underground—cause weathering and erosion, which change the land's surface features and create underground formations.	• Students observe formation of groundwater on a model and then read about the water cycle, and how the depletion of groundwater leads to sinkholes. • Students explain how water, combined with limestone's chemical properties, leads to karst formations. Additionally, students consider how human water consumption affects this system.
Crosscutting concepts	• Cause and Effect • Energy and Matter: Flows, Cycles, and Conservation • Systems and System Models	• Students follow a complicated chain of interactions that begins with rainfall over certain terrain and ends with water consumption and sinkholes. They respond to the following prompt: In your lab, you were asked to guess how limestone and groundwater could be related. Based on the article, describe at least one relationship between limestone and groundwater. • Students use the water cycle to understand sinkholes and see how limited quantities of water affect the karst ecosystem. • Students use models, both physical and in the form of diagrams, to understand the larger system of karst formation, and see how that system interacts with the water cycle.

Chapter 9 *(Continued)*

CCSS Correlations		
Reading standard(s)	• **CCSS.ELA-Literacy.RST.6-8.2.** Determine the central ideas or conclusions of a text; provide an accurate summary of the text distinct from prior knowledge or opinions. • **CCSS.ELA-Literacy.RST.6-8.9.** Compare and contrast the information gained from experiments, simulations, video, or multimedia sources with that gained from reading a text on the same topic.	• *Reading strategy:* Talk your way through it • In this strategy, students practice summarizing the text as they read. • Students use a model and then read about the same topic. They compare the two sources of information to derive conclusions about the effect of the water cycle on karst landscapes.
Writing standard(s)	• **CCSS.ELA-Literacy.WHST.6-8.2.** Write informative/ explanatory texts, including the narration of historical events, scientific procedures/ experiments, or technical processes.	• *Writing prompt:* Late one night in 2011, a UPS store in Georgetown, South Carolina, collapsed into a sinkhole. The store was located over a limestone aquifer that had been stable until the collapse. The owners are suing the South Carolina Department of Transportation because the department had been draining groundwater nearby to allow for the installation of some underground structures. Pretend you are filing a legal brief (report) to support the UPS store's case. Explain how the Department of Transportation's groundwater work might have led to a sinkhole under the UPS store.

Chapter 10

Standard
MS-ESS2. Earth's Systems (*www.nextgenscience.org/msess2-earth-systems*)

Performance expectations
The materials/lessons/activities outlined in this chapter are just one step toward reaching the performance expectations listed below.

MS-ESS2-6. Develop and use a model to describe how unequal heating and rotation of the Earth cause patterns of atmospheric and oceanic circulation that determine regional climates.

Dimension	Element	Matching Student Task or Question From the Activity
Science and engineering practices	• Developing and Using Models	• Students use physical models to look at factors that affect ocean currents. • Students answer questions about how their physical models represent larger systems. • Students read about how new research is challenging an existing digital model of ocean currents.
Disciplinary core ideas	**ESS2.C.** The Roles of Water in Earth's Surface Processes • The complex patterns of the changes and the movement of water in the atmosphere, determined by winds, landforms, and ocean temperatures and currents, are major determinants of local weather patterns. • Variations in density due to variations in temperature and salinity drive a global pattern of interconnected ocean currents.	• Students observe variables (heat and salt) that can affect water density and ocean currents. • Students read about the factors that affect ocean currents and research using floats to observe ocean currents.
Crosscutting concepts	• Systems and System Models	• *Writing prompt:* Scientists often create models to show how they think something works. Do scientists ever change these models? Provide an example from the text to support your answer.
CCSS Correlations		
Reading standard(s)	• **CCSS.ELA-Literacy.RST.6-8.7.** Integrate quantitative or technical information expressed in words in a text with a version of that information expressed visually (e.g., in a flowchart, diagram, model, graph, or table).	• *Reading strategy:* Previewing diagrams and illustrations • *Writing prompt:* Students are asked to compare the simulation to the text to determine why the water layered as it did.
Writing standard(s)	• **CCSS.ELA-Literacy.RST.6-8.9.** Compare and contrast the information gained from experiments, simulations, video, or multimedia sources with that gained from reading a text on the same topic. • **CCSS.ELA-Literacy.WHST.6-8.2.** Write informative/explanatory texts, including the narration of historical events, scientific procedures/experiments, or technical processes.	• *Writing prompt:* Revisit the second model from your lab activity. Using information from the article, explain why the water moved as it did and how this model relates to ocean circulation. Feel free to include diagrams in your explanation. • *Prewriting questions:* Make a quick list of the key ideas you want to include. Number them in the order you will use them. What science words will you want to include? You will be writing about a process that involves cause and effect. What kinds of writing words could help you structure your explanation? (*first, next, then, because, therefore*)

Once Upon an Earth Science Book

Standard

ESS2. Earth's Systems (*www.nextgenscience.org/msess2-earth-systems*)

Performance expectations

The materials/lessons/activities outlined in this chapter are just one step toward reaching the performance expectations listed below.

MS-ESS2-6. Develop and use a model to describe how unequal heating and rotation of the Earth cause patterns of atmospheric and oceanic circulation that determine regional climates.

Dimension	Element	Matching Student Task or Question From the Activity
Science and engineering practices	• Constructing Explanations	• Students use information from observation, from data, and from their reading to explain how wind and currents cause garbage to collect in specific places on Earth.
Disciplinary core ideas	**ESS2.C.** The Roles of Water in Earth's Surface Processes • The complex patterns of the changes and the movement of water in the atmosphere, determined by winds, landforms, and ocean temperatures and currents, are major determinants of local weather patterns. **ESS3.C.** Human Impacts on Earth Systems • Human activities have significantly altered the biosphere, sometimes damaging or destroying natural habitats and causing the extinction of other species. But changes to Earth's environments can have different affects (negative and positive) for different living things.	• Students observe how wind affects water by blowing over water with a straw and watching how the water moves and what happens when it encounters landforms. • Students use maps of ocean currents to explain differences in mean temperatures in cities at similar latitudes and elevation. • Students read about how ocean currents and the Coriolis effect concentrate plastic waste in certain areas of the ocean.
Crosscutting concepts	• Scale, Proportion, and Quantity	• Students observe small models of several variables (the Coriolis effect, wind on water, currents on garbage, etc.) Then they read about how these small variables work together over a large scale to affect global garbage distribution.
CCSS Correlations		
Reading standard(s)	• **CCSS.ELA-Literacy.RST.6-8.1.** Cite specific textual evidence to support analysis of science and technical texts. • **CCSS.ELA-Literacy.RST.6-8.5.** Analyze the structure an author uses to organize a text, including how the major sections contribute to the whole and to an understanding of the topic.	• *Reading strategy:* Identifying Text Signals for Cause and Effect
Writing standard(s)	• **CCSS.ELA-Literacy.WHST.6-8.2.** Write informative/explanatory texts, including the narration of historical events, scientific procedures/experiments, or technical processes. • **CCSS.ELA-Literacy.WHST.6-8.2.c.** Use appropriate and varied transitions to create cohesion and clarify the relationships among ideas and concepts.	• *Writing prompt:* Cargo ships sometimes lose all or part of their load during storms at sea. Suppose a ship loaded with rubber duckies spilled its cargo off the coast of Morocco in North Africa. Describe two locations, other than the coast of Africa, where the ducks might end up and tell why they might end up there. • *Prewriting questions:* (1) Help students locate Morocco on a map showing ocean currents. (2) What science terms will you want to include in your writing? (3) You may want to use writing words related to cause and effect. What kinds of writing words would be helpful? (*because, therefore, so,* etc.)

184

Chapter 12

Standard		
MS-ESS2. Earth's Systems (*www.nextgenscience.org/msess2-earth-systems*)		
Performance expectations		
The materials/lessons/activities outlined in this chapter are just one step toward reaching the performance expectations listed below. **MS-ESS2-5.** Collect data to provide evidence for how the motions and complex interactions of air masses results in changes in weather conditions. **MS-ESS2-6.** Develop and use a model to describe how unequal heating and rotation of the Earth cause patterns of atmospheric and oceanic circulation that determine regional climates.		

Dimension	Element	Matching Student Task or Question From the Activity
Science and engineering practices	• Engaging in Argument From Evidence	• *Writing prompt:* Think back to your lab. Why didn't the balloon pop when it was filled with water? Make a claim, support it with evidence, and explain how the evidence supports your claim.
Disciplinary core ideas	**ESS2.D.** Weather and Climate • The ocean exerts a major influence on weather and climate by absorbing energy from the Sun, releasing it over time, and globally redistributing it through ocean currents.	• Students do two activities to observe water's heat capacity. They read about how this heat capacity affects wind and leads to hurricanes.
Crosscutting concepts	• Systems and System Models	• Students read and write about how heat capacity, sunlight, and wind work together to create different weather patterns depending on latitude. • *The Big Question:* How does water's heat capacity cause the wind to blow at the beach?
CCSS Correlations		
Reading standard(s)	• **CCSS.ELA-Literacy.RST.6-8.1.** Cite specific textual evidence to support analysis of science and technical texts. • **CCSS.ELA-Literacy.RST.6-8.9.** Compare and contrast the information gained from experiments, simulations, video, or multimedia sources with that gained from reading a text on the same topic.	• *Reading strategy:* Identifying text signals for comparisons and contrasts
Writing standard(s)	• **CCSS.ELA-Literacy.WHST.6-8.1.** Write arguments focused on discipline-specific content.	• *Writing prompt:* Think back to your lab. Why didn't the balloon pop when it was filled with water? Make a claim, support it with evidence, and explain how the evidence supports your claim.

Once Upon an Earth Science Book

Chapter 13

Standard		
MS.ESS1. Earth's Place in the Universe (*www.nextgenscience.org/msess1-earth-place-universe*)		

Performance expectations
The materials/lessons/activities outlined in this chapter are just one step toward reaching the performance expectations listed below.

MS-ESS1-3. Analyze and interpret data to determine scale properties of objects in the solar system.

Dimension	Element	Matching Student Task or Question From the Activity
Science and engineering practices	• Developing and Using Models • Analyzing and Interpreting Data	• At each station in the exploration, students use models representing size, distance, composition, and satellites of the planets. • Students look for patterns within their data in order to group planets in a data-based manner.
Disciplinary core ideas	**ESS1.B.** Earth and the Solar System • The solar system consists of the Sun and a collection of objects, including planets, their moons, and asteroids that are held in orbit around the Sun by its gravitational pull on them. • The solar system appears to have formed from a disk of dust and gas, drawn together by gravity.	• Students observe planetary characteristics and compare that information to other data, presenting through reading, which scientists use to group the planets and explain their origins.
Crosscutting concepts	• Patterns • Scale, Proportion, and Quantity • Systems and System Models	• Students analyze data to see patterns that might explain planetary origins. • Students walk through the distance between planets in a model to develop an understanding of the vastness of the universe. They also compare models of the planets to see their relative size. • Students use data to determine why physical models cannot show both distance and size in the solar system accurately.
CCSS Correlations		
Reading standard(s)	• **CCSS.ELA-Literacy.RST.6-8.7.** Integrate quantitative or technical information expressed in words in a text with a version of that information expressed visually (e.g., in a flowchart, diagram, model, graph, or table). • **CCSS.ELA-Literacy.RST.6-8.9.** Compare and contrast the information gained from experiments, simulations, video, or multimedia sources with that gained from reading a text on the same topic.	• *Reading strategy:* Previewing diagrams and illustrations • *Pre-reading prompt:* Tell students to watch for information that would explain why the same planets kept ending up in the same groups during their exploration.
Writing standard(s)	• **CCSS.ELA-Literacy.WHST.6-8.2.** Write informative/explanatory texts, including the narration of historical events, scientific procedures/experiments, or technical processes.	• In another solar system in our galaxy, scientists have found a planet that is a lot like Jupiter. It is large, not very dense, and mostly made of light gases. However, this gas giant orbits very close to its star. Scientists suspect this gas giant's orbit may have gotten smaller at some point. Assuming this solar system developed the same way that ours did, why would scientists suspect that the planet has moved?

Chapter 14

Standard
MS-ESS1. Earth's Place in the Universe (*www.nextgenscience.org/msess1-Earth-place-universe*)

Performance expectations
The materials/lessons/activities outlined in this chapter are just one step toward reaching the performance expectations listed below.

MS-ESS1-1. Develop and use a model of the Earth-Sun-Moon system to describe the cyclic patterns of lunar phases, eclipses of the Sun and Moon, and seasons.

Dimension	Element	Matching Student Task or Question From the Activity
Science and engineering practices	• Developing and Using Models	• Students create a model of the Earth and Sun to explain how differing seasons could take place for two cities at similar latitudes north and south of the equator. • Then students use their model to explain data about light and dark periods and the poles.
Disciplinary core ideas	**ESS1.A.** The Universe and Its Stars • Patterns of the apparent motion of the Sun, the Moon, and stars in the sky can be observed, described, predicted, and explained with models. **ESS1.B.** Earth and the Solar System • This model of the solar system can explain eclipses of the Sun and the Moon. Earth's spin axis is fixed in direction over the short term but tilted relative to its orbit around the Sun. The seasons are a result of that tilt and are caused by the differential intensity of sunlight on different areas of Earth across the year.	• Students create a model of the Earth and Sun to explain how differing seasons could take place for two cities at similar latitudes north and south of the equator. • Students use a flashlight and graph paper to determine the difference in energy input for direct and indirect sunlight.
Crosscutting concepts	• Systems and System Models	• Students develop a model to fit one set of data (the seasons in two cities) and then use that model to explain extended patterns of light and dark at the North Pole. • Students read about how the same model explains seasons on other planets, such as Uranus.
CCSS Correlations		
Reading standard(s)	• **CCSS.ELA-Literacy.RST.6-8.7.** Integrate quantitative or technical information expressed in words in a text with a version of that information expressed visually (e.g., in a flowchart, diagram, model, graph, or table).	• *Reading Strategy.* Previewing diagrams and illustrations
Writing standard(s)	• **CCSS.ELA-Literacy.WHST.6-8.1.** Write arguments focused on discipline-specific content. • **CCSS.ELA-Literacy.WHST.6-8.1.a.** Introduce claim(s) about a topic or issue, acknowledge and distinguish the claim(s) from alternate or opposing claims, and organize the reasons and evidence logically.	• *Writing prompt.* A classmate incorrectly claims that winter is when Earth is farthest away from the Sun in its orbit, and summer happens when Earth is closest to the Sun. Make a correct claim about what causes season, and support it with evidence and reasons from your lab and the article. In your response, provide evidence that rebuts your classmates' claim. • *Prewriting questions.* What is a rebuttal? How can you include a rebuttal when writing about a claim? What kinds of writing words might you use in a rebuttal? (*If ... then* statements, *however,* and *therefore* are useful words for rebuttals.)

Chapter 15

Standard		
MS-ESS3. Earth and Human Activity (*www.nextgenscience.org/msess3-Earth-human-activity*)		

Performance expectations
The materials/lessons/activities outlined in this chapter are just one step toward reaching the performance expectations listed below.

MS-ESS3-3. Apply scientific principles to design a method for monitoring and minimizing a human effect on the environment.
MS-ESS3-4. Construct an argument supported by evidence for how increases in human population and per-capita consumption of natural resources affect Earth's systems.

Dimension	Element	Matching Student Task or Question From the Activity
Science and engineering practices	• Planning and Carrying Out Investigations • Obtaining, Evaluating, and Communicating Information	• Students design a method and collect data to decide which hair dryer is best. • Students read and evaluate an argument on government policies for energy efficiency.
Disciplinary core ideas	**ESS3.A.** Natural Resources • Humans depend on Earth's land, ocean, atmosphere, and biosphere for many different resources. Minerals, freshwater, and biosphere resources are limited, and many are not renewable or replaceable over human lifetimes. These resources are distributed unevenly around the planet as a result of past geologic processes. **ESS3.C.** Human Impacts on Earth Systems • Typically as human populations and per-capita consumption of natural resources increase, so do the negative impacts on Earth unless the activities and technologies involved are engineered otherwise.	• Students consider energy efficiency in light of the limits of fossil fuel (and renewable) energy sources. • Students answer questions about a chart that depicts the population and energy use in a variety of countries.
Crosscutting concepts	• Patterns	• Students both generate charts and use charts to organize their evidence and thinking on the topic of energy efficiency.
CCSS Correlations		
Reading standard(s)	• **CCSS.ELA-Literacy.RST.6-8.2.** Determine the central ideas or conclusions of a text; provide an accurate summary of the text distinct from prior knowledge or opinions. • **CCSS.ELA-Literacy.RST.6-8.5.** Analyze the structure an author uses to organize a text, including how the major sections contribute to the whole and to an understanding of the topic. • **CCSS.ELA-Literacy.RST.6-8.6.** Analyze the author's purpose in providing an explanation, describing a procedure, or discussing an experiment in a text.	• *Reading strategy:* Evaluating persuasive science writing
Writing standard(s)	• **CCSS.ELA-Literacy.WHST.6-8.9.** Draw evidence from informational texts to support analysis, reflection, and research.	• *Prompt:* What parts of the author's argument in "Mandate Energy Efficiency!" do you accept? Did the author support her position well? Why or why not? What other ideas should she have considered?

About the Author

Jodi Wheeler-Toppen is the author of a dozen books for children and teachers published through National Geographic Kids, Capstone Press, and the National Science Teachers Association (NSTA Press). She holds a PhD in science education from the University of Georgia and lives in Atlanta, Georgia, with her family. She can be found on the web at *www.OnceUponAScienceBook.com*.

Image Credits

Chapter 4

p. 35: Jodi Wheeler-Toppen

p. 36: Pixabay, Public domain. *https://pixabay.com/en/collie-dog-pet-animal-385088.*

p. 37: NSTA Press

pp. 41–43: Joseph Peterson, PLoS ONE, Public domain. *http://journals.plos.org/plosone/article?id=10.1371/journal.pone.0068620.*

Chapter 5

p. 47: Jodi Wheeler-Toppen

p. 51: Vermont Agency of Transportation, Public domain. *www.anr.state.vt.us/anr/climatechange/Images/Irene_InstreamWork.jpg.*

p. 51: Brigitte Werner, Pixabay, Public domain. *https://pixabay.com/en/pink-sand-dunes-utah-usa-desert-65310.*

p. 52: Pixabay, Public domain. *https://pixabay.com/en/rice-terraces-rice-fields-164410.*

p. 53: Jodi Wheeler-Toppen

Chapter 6

pp. 59–61: NSTA Press

p. 62: Naturmuseum Senckenberg, Wikimedia Commons, Public domain. *https://commons.wikimedia.org/wiki/File:Polar_coat_-_Alfred_Wegener_-_Naturmuseum_Senckenberg_-_DSC02110.JPG.*

p. 63: U.S. Geological Survey, Public domain.

p. 64: NSTA Press

Chapter 7

p. 70: Jodi Wheeler-Toppen

p. 71: Apokryltaros, Wikimedia Commons, CC BY-SA 3.0. *https://en.wikipedia.org/wiki/User:Apokryltaros#/media/File:Anomalocaris_saron.jpg.*

p. 71: Danny Nicholson, Flickr, CC BY-ND 2.0. *www.flickr.com/photos/dannynic/3845056251/in/faves-15370532@N02.*

p. 72: Edna Winti, Flickr, CC BY 2.0. *www.flickr.com/photos/ednawinti/9576213233.*

p. 72: Mark A. Wilson, Wikimedia Commons, Public domain. *https://commons.wikimedia.org/wiki/File:Cambrian_Trilobite_Olenoides_Mt._Stephen.jpg.*

p. 74: National Park Service, Public domain. *www.nature.nps.gov/Geology/parks/grca/age/image_popup/yardstick.htm.*

Chapter 8

p. 80: U.S. Geological Survey, Public domain. *http://library.usgs.gov/photo/#/item/51ddb5dde4b0f72b4471fa30.*

p. 81: U.S. Geological Survey, Public domain. *http://gallery.usgs.gov/photos/01_08_2013_x16Fw32VUp_01_08_2013_0#.VcofTq1UXn1.*

p. 81: U.S. Geological Survey, Public domain. *http://library.usgs.gov/photo/#/item/51dc23fce4b0f81004b78e66.*

p. 82: NSTA Press

p. 84: Jodi Wheeler-Toppen

Chapter 9

p. 90: NSTA Press

p. 92: U.S. Geological Survey, Public domain. *http://library.usgs.gov/photo/#/item/51ddc5cae4b0f72b44720f4b.*

p. 93: Geological Survey of Canada, with permission. *https://web.viu.ca/geoscape/images/karst_cave.jpg.*

p. 94: U.S. Geological Survey, ID: DI00000001142201, Public domain. *www.usgs.gov/laws/info_policies.html.*

p. 96: NSTA Press

Chapter 10

pp. 99–104: Jodi Wheeler-Toppen

p. 105: Danski114, Wikimedia Commons, CC-BY-SA-3.0. *https://commons.wikimedia.org/wiki/File:Ice_XI_View_along_c_axis.png.*

p. 105: NSTA Press

p. 105: Susan Lozier, Nicholas School of the Environment, Duke University, with permission. *www.flickr.com/photos/nicholasschoolatduke/3548702245/in/photolist-odwHVy-p5krzA-suuSFs-7Yjhg2-7Yjh8p-7Ynwkm-7YjhnR-pcB6Un-6pA2xt-6pEoFf-ecUAFX-e3dmt8.*

p. 107: NSTA Press

Chapter 11

pp. 111–113: Jodi Wheeler-Toppen

p. 117: Wikimedia Commons, Public domain. *https://commons.wikimedia.org/wiki/File:Worldmap_northern.svg.*

p. 118: Wikimedia Commons, Public domain. *https://commons.wikimedia.org/wiki/File:Worldmap_southern.svg.*

p. 119: Jodi Wheeler-Toppen

p. 119: Michael Pidwirney, Wikimedia Commons, with permission. *https://commons.wikimedia.org/wiki/File:Corrientes-oceanicas.png.*

p. 120: NSTA Press

p. 121: National Oceanic and Atmospheric Administration, Marine Debris Program, Public domain. *http://marinedebris.noaa.gov/info/patch.html.*

p. 121: Algalita Marine Research and Education, with permission.

p. 121: Starr Environmental, CC-BY 3.0. *www.starrenvironmental.com/images/images/plants/full/13/09/starr-130910-0711.jpg.*

Chapter 12

p. 130: Wikimedia Commons, Public domain. *https://commons.wikimedia. org/wiki/File:Wind_circulation.svg.*

p. 130: NSTA Press

p. 131: Learn NC, Public domain. *www. learnnc.org/lp/multimedia/2367.*

p. 131: National Oceanic and Atmospheric Administration, Public domain. *www. nnvl.noaa.gov/MediaDetail2. php?MediaID=1725&MediaTypeID=1.*

Chapter 13

p. 137: Karen Kraus, with permission.

pp. 140–141: NSTA Press

p. 148: NASA, Public domain. *www. nasa.gov/mission_pages/juno/launch/ Juno_launchpreview.html.*

p. 148: NSTA Press

p. 149: International Astronomical Union/Martin Kornmesser, International Astronomical Union, CC-BY-3.0. *www.iau.org/public/images/detail/ iau0601a.*

Chapter 14

pp. 153–154: Jodi Wheeler-Toppen

p. 159: NSTA Press

p. 160: Dennis Cain, National Atmospheric and Oceanic Administration, Public domain.

pp. 160–161: NSTA Press

Chapter 15

p. 170: Adam Buchbinder, Flickr, CC BY-SA 2.0. *www.flickr.com/ photos/grendelkhan/106679350/in/ photolist-aqL49.*

p. 170: Pixabay, Public domain. *https:// pixabay.com/en/blow-dryer-blow-drier- hair-blower-311549.*

Index

Page numbers in **boldface** type refer to tables or figures.